Real-Life Math

FRACTIONS, RATIOS, AND RATES

SECOND EDITION

WALCH PUBLISHING

To Lori

The classroom teacher may reproduce materials in this book for classroom use only.
The reproduction of any part for an entire school or school system is strictly prohibited.
No part of this publication may be transmitted, stored, or recorded in any form
without written permission from the publisher.

1 2 3 4 5 6 7 8 9 10

ISBN 978-0-8251-6320-3

Copyright © 1998, 2007

J. Weston Walch, Publisher

P. O. Box 658 • Portland, Maine 04104-0658

www.walch.com

Printed in the United States of America

Table of Contents

Table of Contents

Rates with Exponents

Ratios

How to Use This Series

The *Real-Life Math* series is a collection of activities designed to put math into the context of real-world settings. This series contains math appropriate for pre-algebra students all the way up to pre-calculus students. Problems can be used as reminders of old skills in new contexts, as an opportunity to show how a particular skill is used, or as an enrichment activity for stronger students. Because this is a collection of reproducible activities, you may make as many copies of each activity as you wish.

Please be aware that this collection does not and cannot replace teacher supervision. Although formulas are often given on the student page, this does not replace teacher instruction on the subjects to be covered. Teaching notes include extension suggestions, some of which may involve the use of outside experts. If it is not possible to get these presenters to come to your classroom, it may be desirable to have individual students contact them.

We have found a significant number of real-world settings for this collection, but it is not a complete list. Let your imagination go, and use your own experience or the experience of your students to create similar opportunities for contextual study.

Foreword

As a mathematician, I have to admit that I always used to greet the question, "When are we ever going to use this?" with some annoyance. It was obvious to me that students needed the material to do well on my tests and maybe in future classes. Admittedly, that doesn't give students, unless they really just love math, much on which to hang their hats. With greater experience, I began answering the question with esoteric phrases about how well math trains one's mind and why training the mind is the highest goal of a good education. Still, some students stare at me blankly, trying to find the "real" meaning of their math voyage.

Well, we really DO use math every day. Yes, sometimes it is just to balance a checkbook or to make change, but in an incredible variety of professional and personal settings, we use math skills that were drilled into us without the slightest hint of a context. Once I started to think about all the areas of life where math and mathematical thought were central, I started having fun. I have talked with stockbrokers, restaurateurs, mechanics, haberdashers, contractors, baseball statisticians, bankers, carpet salespeople, and grocers to learn about how they use math each day. I hope that these activities will be as much fun for your students and provide them with as much contextual background as they have for me. As a teacher, you will be amazed by how open professionals in other fields are to helping your students extend their understanding and grounding in learning math.

—Tom Campbell

© 2007 Walch Publishing

Real-Life Math: Fractions, Ratios, and Rates

1. A Construction Site

Goal

To use adding and subtracting fractions in a construction setting

Context

Many of us have to work with at least some sense of fractions. Most rulers and measuring sticks are scaled in fractional parts of inches, not decimal parts. Consequently, a construction or refurbishment site is a terrific source of fraction problems.

Teaching Notes

Many of us learned fractions for the first time by talking about cutting up pies and cakes. Almost immediately afterward, we were manipulating fractions without a context. The question "What is $1/7 + 3/5$?" has an answer, but not one we can place in context. Practicing addition and subtraction of fractions is easier for students if they have a good sense of what the fraction is "part of."

Extension Activity

Ask students to measure out and draw a scale drawing of a room in their house. They should identify the location of furniture, rugs, and so on. Scale drawings make use of both ratios and fractions.

Answers

1. 9 feet, $5\,^{15}/_{16}$ inches

2. 12 feet, 11 inches $\times$ 10 feet, $^1/_2$ inch

3. 4 feet, $1\,^3/_4$ inches

4. $4\,^7/_{16}$ inches

5. $5\,^1/_4$ inches $\times$ $3\,^1/_2$ inches

6. 25 feet, $10\,^1/_2$ inches

7. 2 feet, $11\,^3/_8$ inches

8. $95\,^1/_8$ inches (7 feet, $11\,^1/_8$ inches) $\times$ $51\,^1/_4$ inches (4 feet, $3\,^1/_4$ inches)

1. A Construction Site

Rulers and measuring tapes are often divided into inches. Inches are sometimes broken into halves, quarters, eighths, tenths, or sixteenths depending on the measurer. The inches can also be grouped in sets of 12 into feet. Most measurements in the construction business are given in feet, inches, and fractions of inches.

Brian McDonald has a general contracting business. He and his employees always measure carefully to make sure they cut their materials correctly. On a recent job site, they encountered the following situations. Give Brian's crew a hand with their measurements by answering the questions below.

1. Brian is putting in a countertop. He is going to put the countertop flush against the wall on one end and trim it with a board that is $9/16$ inch thick on the other end. If Brian wants the counter to end exactly 9 feet, $6\frac{1}{2}$ inches from the wall, how long should he cut the countertop?

2. Brian has hired Matteen to build the deck. Matteen can just fit a deck that is 13 feet, 6 inches along the side of the house by 10 feet 4 inches out from the house. The customer wants a railing; however, that will be $3\frac{1}{2}$ inches wide, running around the three sides not attached to the house. What will be the dimensions of the "livable space" on the deck?

3. Brian has hired Sarah to build a square window frame out of wood that is $3/4$ inch thick. The dimensions need to be 4 feet, $3\frac{1}{4}$ inches square. If Sarah cuts the two sides of the square to be 4 feet, $3\frac{1}{4}$ inches long, how long should she cut the top and bottom pieces to fit between these two sides?

(continued)

1. A Construction Site

4. Lori has been hired to build the frame and outer walls of the house. The exterior walls will be made of $3/8$-inch drywall, $3\frac{1}{2}$-inch-wide studs, $1/4$-inch-thick exterior plywood, and $5/16$-inch-thick siding. How thick are the new walls altogether?

5. Rayhan has been hired to build a rectangular box out of $3/8$-inch-thick wood as a curio shelf. He wants the box to be 6 inches long and $4\frac{1}{4}$ inches wide. If he insets the bottom into the box, to what dimensions should he cut the piece for the bottom?

6. The house is 26 feet, $4\frac{3}{4}$ inches wide on the outside. Brian wants to cut his clapboards to fit the sides, but he knows that the corner trim covers $3\frac{1}{8}$ inches in from the edge on each end. How long should he cut the clapboards?

7. Concrete will be poured into forms at the job site tomorrow. The wall to be poured is to be 15 feet, $5\frac{3}{8}$ inches long. Rayhan and Brian already have a form that is 12 feet, 6 inches long. How long should they make an additional form to make the wall the right length?

8. Brian has asked Matteen to build a doorway for an existing door. Matteen knows that the door is $41\frac{3}{4}$ inches wide and $90\frac{3}{8}$ inches tall. He knows the jamb will be $4\frac{3}{4}$ inches thick on the top and two sides of the doorway. What are the dimensions of the hole he should create in the wall to accommodate the doorjamb and the door?

2. Selling Carpet

Goal

To learn about fractions and areas in the context of selling carpeting by the square yard

Context

Most floor coverings are sold by the square yard. Carpeting in particular is priced this way. The shape of the original roll of carpet is not as important as the area of the actual room. Students may have lived in homes where carpeting was installed or may be interested in a sales position, such as that of a carpet salesperson.

Teaching Notes

Students need to be familiar with conversions among yards, feet, and inches, and with how to calculate the area of nonrectangular polygons with right-angle corners. The easiest way to do this is to break the shapes into consecutive rectangles, multiply sides to find their various areas, and then add these various areas.

Extension Activity

Ask students to measure several rooms in the school or at home in order to find the exact floor area of the space.

Answers

1. The Winstons should be charged $265.17.

2. The Khins should be charged $468.04.

3. The Traftons should be charged $353.05.

4. The Cowens should be charged $683.26.

5. The Taylors should be charged $1188.24.

2. Selling Carpet

Kevin's Carpet World has been installing office carpeting for years. They are now trying to break into the home carpeting business. When customers come into the salesroom, Kevin asks them to bring an accurate floor plan of the room(s) to be carpeted. The staff can then calculate the square yardage of the floor to determine the price of the carpeting. Kevin's prices include a 6-pound pad and installation, so customers have only one price to consider.

For each situation below, help Kevin determine the price of the carpet installation.

1. The Winston family is installing a playroom in their basement. They like the Berber carpet that is on sale for $12.99 a square yard. What price should be charged assuming the following floor plan is accurate?

    Total cost: _____

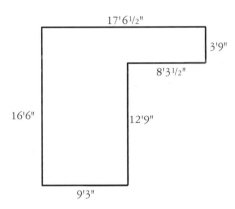

2. The Khin family is carpeting their living room. They have chosen a deep-pile carpet that sells for $24.95 a square yard. How much should they be charged based on the following floor plan?

    Total cost: _____

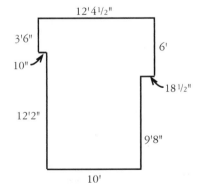

(continued)

5   *Real-Life Math: Fractions, Ratios, and Rates*

2. Selling Carpet

3. The Trafton family wants to install outdoor carpeting around their in-ground pool. The outdoor carpeting they have selected is $8.99 a square yard. How much should they be charged based on the following plan of the pool area?

    Total cost: _____

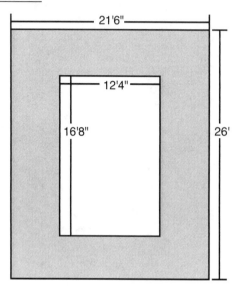

4. The Cowen family is carpeting their dining room. They have chosen a low-pile carpet that sells for $18.75 a square yard. How much should they be charged based on the following floor plan?

    Total cost: _____

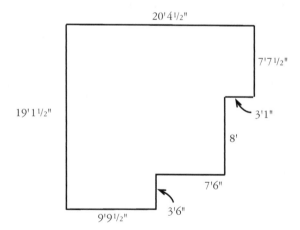

(continued)

2. Selling Carpet

5. The Taylor family is carpeting their bedrooms. They have chosen a soft carpet that sells for $18.15 a square yard. How much should they be charged based on the following floor plans?

   Total cost: _____

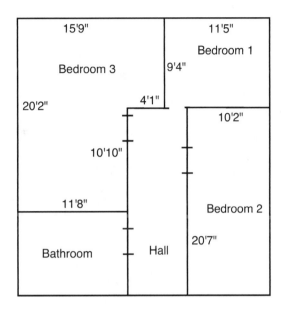

3. Sports Standings

Goal

To be able to read and create a chart of the standings in a sports conference or division

Context

Sports standings are not always as easy to figure out as they might appear. Placement in the table, with first place on the top, is usually based on "winning percentage." The GB, or "games behind" column contains either a whole number or a mixed number including $1/2$. Sometimes the first-place team is "behind" the second-place team because they have not played the same number of games.

Teaching Notes

Two columns in the following baseball and basketball standings are of interest (neither sport has ties, which change this procedure): Winning Percentage "PCT," which is the number of wins divided by the number of games played so far, and Games Behind "GB," which is the number of wins by Team A minus the number of wins by Team B, plus the number of losses by Team B minus the number of losses by Team A. This sum is divided by 2. If this result is positive, it is the number of games B is behind A, and if it is negative, the number of games A is behind B.

Extension Activities

It is almost always baseball or basketball season. Have students use the newspaper to find standings and check their accuracy, or have students calculate the standings for your teams' leagues.

Answers

1. National League Final Standings 1962

Team	Wins	Losses	PCT	GB
St. Louis	93	69	.574	—
Cincinnati	92	70	.568	1
Philadelphia	92	70	.568	1
San Francisco	90	72	.556	3
Milwaukee	88	74	.543	5
Los Angeles	80	82	.494	13
Pittsburgh	80	82	.494	13
Chicago	76	86	.469	17
Houston	66	96	.407	27
New York	53	109	.327	40

(continued)

3. Sports Standings

2. Eastern Conference Standings for the
    NBA as of March 19, 2006

Atlantic Division				
Team	Wins	Losses	PCT	GB
New Jersey	36	28	.563	—
Philadelphia	31	34	.477	5.5
Boston	27	39	.409	10
Toronto	24	42	.364	13
New York	19	45	.297	17

Central Division				
Team	Wins	Losses	PCT	GB
Detroit	52	13	.800	—
Cleveland	37	29	.561	15.5
Indiana	33	30	.524	18
Milwaukee	33	33	.500	19.5
Chicago	29	37	.439	23.5

Southeast Division				
Team	Wins	Losses	PCT	GB
Miami	44	21	.677	—
Washington	33	31	.516	10.5
Orlando	24	41	.369	20
Atlanta	20	43	.317	23
Charlotte	18	49	.269	27

(continued)

3. Sports Standings

3.

Team	PCT	Reason for Invitation
1. Detroit	.800	First place in Central Division
2. Miami	.677	First place in Southeast Division
3. New Jersey	.563	First place in Atlantic Division
4. Cleveland	.561	Best percentage among non-leaders of division
5. Indiana	.524	Second best percentage among non-leaders of division
6. Washington	.516	Third best percentage among non-leaders of division
7. Milwaukee	.500	Fourth best percentage among non-leaders of division
8. Philadelphia	.477	Fifth best percentage among non-leaders of division

4.   Chicago is in ninth place with a
.439 winning percentage. They are
2.5 games behind Philadelphia.

© 2007 Walch Publishing   *Real-Life Math: Fractions, Ratios, and Rates*

3. Sports Standings

Newspapers print the standings of various sports leagues daily. The team in first place is usually listed at the top. "First place" status is based on best "winning percentage," but "games behind" can also be used. Jose is a sports editor for the *Sunville Times*. He has to check the standings each day for accuracy (whether they are current or outdated).

Jose computes the winning percentage (PCT) by dividing the number of wins a team has by the number of games they have played. Jose then determines the games behind (GB) for each team. He subtracts a team's wins from the first place team's wins, then adds the team's losses to this difference. The next step is to subtract the first place team's losses. The resulting number is divided by 2.

Help Jose with his job by filling in the PCT and the GB fields in the charts below.

1. National League Final Standings 1962

Team	Wins	Losses	PCT	GB
St. Louis	93	69		
Cincinnati	92	70		
Philadelphia	92	70		
San Francisco	90	72		
Milwaukee	88	74		
Los Angeles	80	82		
Pittsburgh	80	82		
Chicago	76	86		
Houston	66	96		
New York	53	109		

2. Eastern Conference Standings for the NBA as of March 19, 2006

Atlantic Division				
Team	Wins	Losses	PCT	GB
New Jersey	36	28		
Philadelphia	31	34		
Boston	27	39		
Toronto	24	42		
New York	19	45		

(continued)

3. Sports Standings

Southeast Division				
Team	Wins	Losses	PCT	GB
Miami	44	21		
Washington	33	31		
Orlando	24	41		
Atlanta	20	43		
Charlotte	18	49		

Central Division				
Team	Wins	Losses	PCT	GB
Detroit	52	13		
Cleveland	37	29		
Indiana	33	30		
Milwaukee	33	33		
Chicago	29	37		

3. Now that you have calculated all the winning percentages and how far each team is behind in its division, determine who is in the lead to be a playoff team. Each division winner goes to the playoffs, and the five teams that are not division winners but have the best winning percentages are invited to the playoffs as well. Which eight teams would be invited to the playoffs based on these standings? Complete the chart on the next page.

(continued)

3. Sports Standings

Team	PCT	Reason for Invitation
1.		
2.		
3.		
4.		
5.		
6.		
7.		
8.		

4.   Which team is sitting in ninth place, and how far from being in the playoffs are they?

4. Cooking More Chili

Goal

To practice multiplication of fractions and conversion of measurements in the context of adjusting a recipe

Context

Restaurants and cooks frequently want to adjust a given recipe to produce enough servings for their needs. They have to multiply or divide the given measures by a factor that will give them enough servings.

Teaching Notes

Most recipe measures are given in mixed numbers. Adjusting and converting them to change the number of servings produced requires the multiplication or division of mixed numbers. Conversion of measurements (such as 16 tablespoons = 1 cup and 6 teaspoons = 1 ounce) requires that students recognize the conversion factors and multiply or divide accurately.

Extension Activities

- Have students bring in favorite recipes and rework them for a different number of servings. They could also use recipes found on the Internet.

- Make a version of the converted recipe. Just sauté the chopped vegetables in the oil, add all the tomato ingredients, spices, cashews, and raisins, and simmer over low heat for 90 minutes. Stir in vinegar just before serving.

Answers

onions: $8\frac{1}{4}$ pounds

green peppers: $2\frac{3}{4}$ pounds

carrots: $2\frac{3}{4}$ pounds

celery: 22 stalks

vegetable oil: 1 cup + $\frac{1}{2}$ tablespoon

garlic: 11 cloves

canned tomatoes: $5\frac{1}{2}$ quarts

kidney beans: $5\frac{1}{2}$-pound can

tomato sauce: $115\frac{1}{2}$ ounces

tomato paste: 55 ounces

basil, cumin, salt: $\frac{1}{2}$ cup + $\frac{3}{4}$ teaspoons (each)

oregano: $\frac{1}{2}$ cup + 1 tablespoon + $\frac{1}{2}$ teaspoon

chili powder: $\frac{3}{4}$ cup + $1\frac{3}{4}$ tablespoons

bay leaves: 11

black pepper: 2 tablespoons + $\frac{7}{8}$ teaspoon

cashews: $2\frac{3}{4}$ pounds

raisins: $5\frac{1}{2}$ cups

vinegar: $1\frac{3}{8}$ cups

4. Cooking More Chili

Rachel and Jack have a bagel bakery and restaurant. They have decided to create a new vegetarian chili for their customers. In testing recipes, they make small amounts until they find one they like. In cooking for the restaurant, they will use a much bigger pot. The pot they have is $5\frac{1}{2}$ times as big as the tester pot. They have decided on the recipe below. Use this recipe to create the version they should give to their cook. The conversion tables that follow the recipe will help you with your calculations.

Test Recipe Amounts	Ingredient	Full Recipe Amounts
$1\frac{1}{2}$ pounds	chopped onions	
$\frac{1}{2}$ pound	chopped green pepper	
$\frac{1}{2}$ pound	carrots	
4 stalks	chopped celery	
3 tablespoons	vegetable oil	
2 large cloves	garlic	
1 quart	canned tomatoes	
16-ounce can	kidney beans	
21 ounces	tomato sauce	
10 ounces	tomato paste	
$4\frac{1}{2}$ teaspoons	basil	
$1\frac{1}{2}$ tablespoons	cumin	
5 teaspoons	oregano	
$1\frac{1}{2}$ tablespoons	salt	
$2\frac{1}{2}$ tablespoons	chili powder	
2	bay leaves	
$1\frac{1}{4}$ teaspoons	ground black pepper	
$\frac{1}{2}$ pound	raw cashew pieces	
1 cup	raisins	
$\frac{1}{4}$ cup	cider vinegar	

(continued)

4. Cooking More Chili

Volume Conversion Table

1 gallon = 4 quarts = 8 pints = 16 cups = 128 fluid ounces
1 cup = 16 tablespoons = 48 teaspoons
1 fluid ounce = 2 tablespoons = 6 teaspoons
1 tablespoon = 3 teaspoons
1 pound = 16 ounces
1 liter = 1 cubic decimeter

5. Baking Fewer Cookies

Goal

To practice division of fractions and conversion of measurements in the context of adjusting a recipe

Context

Restaurants and cooks frequently want to adjust a given recipe to produce enough servings for their needs. They have to multiply or divide the given measures by a factor that will give them enough servings.

Teaching Notes

Most recipe measures are given in mixed numbers. Adjusting and converting them to change the number of servings produced requires the multiplication or division of mixed numbers. Conversion of measurements (such as 16 tablespoons = 1 cup and 6 teaspoons = 1 ounce) requires that students recognize the conversion factors and multiply or divide accurately.

Extension Activities

- Have students bring in favorite recipes and rework them for a different number of servings. They could also use recipes found on the Internet.

- Make a version of the converted recipe; it's real, and it's delicious.

Answers

butter: $\frac{1}{2}$ cup

flour: 1 cup

baking soda: $\frac{1}{2}$ teaspoon

sugar: $\frac{1}{2}$ cup

oatmeal: $1\frac{1}{4}$ cups

chocolate chips: 6 ounces

brown sugar: $\frac{1}{2}$ cup

salt: $\frac{1}{2}$ teaspoon

grated chocolate bar: 2 ounces

eggs: 1

baking powder: $\frac{1}{2}$ teaspoon

vanilla: $\frac{1}{2}$ teaspoon

chopped walnuts: $\frac{7}{8}$ cup

Name _____ Date _____

5. Baking Fewer Cookies

Kim and Hannah run a bakery. They have decided to put their recipe for chocolate-chip cookies on their web site. When they bake cookies, they make a large batch of 10 dozen cookies at a time, but their web site visitors will probably want to make fewer cookies in one batch. Use the following list of ingredients to create the smaller recipe for 30 cookies. The conversion tables that follow the recipe will help you with your calculations. If necessary, round to the nearest teaspoon or cup.

Bakery Batch Amounts	Ingredient	Web Site Recipe Amounts
2 cups	butter	
4 cups	flour	
1 1/2 teaspoons	baking soda	
1 3/4 cups	sugar	
5 cups	oatmeal (blended fine)	
24 ounces	chocolate chips	
2 1/4 cups	brown sugar	
1 1/2 teaspoons	salt	
8 ounces	grated chocolate bar	
4	eggs	
2 1/4 teaspoons	baking powder	
2 teaspoons	vanilla	
3 1/4 cups	chopped walnuts	

Volume Conversion Table

1 gallon = 4 quarts = 8 pints = 16 cups = 128 fluid ounces
1 cup = 16 tablespoons = 48 teaspoons
1 fluid ounce = 2 tablespoons = 6 teaspoons
1 tablespoon = 3 teaspoons
1 pound = 16 ounces
1 liter = 1 cubic decimeter

6. Miles Per Hour

Goal

To learn how to calculate the familiar rate of speed "miles per hour" (mph) from other measures of time, distance, and speed

Context

Although mph is generally used on roads and highways, we also use it to describe the speed at which many objects travel.

Teaching Notes

The rate mph is used to measure speed, but seldom does it mean exactly 1 mile or exactly 1 hour. By using conversion factors, students can convert other speeds to miles per hour. Miles per hour is usually a measure of average speed. If a sprinter runs 100 yards in 9.9 seconds, we calculate that he has gone 20.7 mph. However, since such races are run from a standing start, the sprinter probably reached a top speed of significantly more than 25 mph at some point during the race. Without a radar gun, we can calculate only the runner's average speed for the distance.

Extension Activities

- Have students look up a record from the Olympics, where the time as well as the distance run, skated, skied, or ridden is given. Have them calculate the average speed of the racer.

- Allow students to conjecture about why average speed rises from short- to medium-length races and then falls again for longer races. (The rise is probably due to the impact of how long the racer takes to attain top speed and the falloff due to stamina.)

Answers

Answers may vary slightly.

1. Seabiscuit's average speed was 37.1 mph for the race.

2. Meagher averaged 3.9 mph for the race.

3. Griffith Joyner averaged 21 mph for the race.

4. Johnson's pitch takes .41 seconds to get to the plate.

5. The linebacker averaged 17.8 mph for the 40-yard dash.

6. Gordon's race lasted 3.66 hours, or 3 hours, 39 minutes, 27.5 seconds.

7. Paolo Bettini's average speed for the race was 24.5 mph.

8. Joey Cheek's average speed for the race was 32.1 mph.

6. Miles Per Hour

Keisha is fascinated by the speed athletes can achieve. She has investigated many sports and enjoys using times and distances to calculate speed. Keisha has carefully researched a variety of races and recorded speeds in sports in an effort to calculate time, distance, or speed.

To determine these rates, Keisha uses the formula $d/t = s$ (distance divided by time equals speed). She likes to express her answers in miles per hour (mph), so she uses the conversion charts below to change her data into miles and hours before she does her computation. Help Keisha with these calculations by answering the questions that follow.

Conversion Factors

English Length	Metric Length
1 inch	2.54 centimeters
.394 inches	1 centimeter
1 foot	30.48 centimeters
.033 feet	1 centimeter
1 yard	9144 meters
1.094 yards	1 meter
1 mile	1.609 kilometers
.622 miles	1 kilometer

Volume	
33.8 ounces	1 liter
1 ounce	.03 liters

Temperature

Fahrenheit	Celsius
Fahrenheit degrees	$\frac{9}{5} \times$ (Celsius degrees) + 32

(continued)

6. Miles Per Hour

Comparative Measures

Distance			
1 mile	= 1,760 yards	= 5280 feet	= 63,360 inches
1 yard	= 3 feet	= 36 inches	
1 foot	= 12 inches		
1 kilometer	= 1000 meters	= 100,000 centimeters	= 1,000,000 millimeters
1 meter	= 100 centimeters		
1 centimeter	= 10 millimeters		

1. On November 1, 1938, Seabiscuit and War Admiral, two of the greatest race horses of the 1930s, were pitted against each other in a match race at Pimlico Race Track. Seabiscuit won in an upset, completing the 1.1875 ($1\,^3/_{16}$)-mile course in 1 minute and 56.6 seconds (a track record at the time). What was Seabiscuit's average speed in mph (to the nearest tenth) for the entire race?

2. In 1981, Mary T. Meagher of the United States swam 100 meters (using the butterfly stroke) in 57.93 seconds. What was her average speed in mph for the race?

3. In the 1988 Olympics in Seoul, Korea, Florence Griffith Joyner set a world record by running 200 meters in 21.34 seconds. What was her average speed in mph for the race?

(continued)

6. Miles Per Hour

4. Randy Johnson is one of the best pitchers of the current baseball era. His pitches have been clocked by radar at 101 mph. The ball travels 60 feet, 6 inches from the pitcher's mound to home plate. How long does the ball take to get to the batter?

5. Football scouts often time players in the 40-yard dash, which is a measure of both quickness and speed. You might hear that a linebacker "runs a 4.6 40," which means he ran the 40-yard dash in 4.6 seconds. What was his average speed in mph for the dash?

6. Jeff Gordon, a top NASCAR driver, averaged 136.7 mph for a 500-mile race. How much time passed from the start of the race to the end?

7. In the 2004 Olympics in Athens, Greece, Italian cyclist Paolo Bettini won the 224.4 km bike race, finishing with a winning time of 5 hours, 41 minutes, and 44 seconds. What was Bettini's average speed in miles per hour (to the nearest tenth) for the entire race?

8. In the 2006 Olympics in Torino, Italy, American Joey Cheek won the 500-meter speed skating gold medal. In the first heat of the race, he skated 500 meters in 34.82 seconds. What was Cheek's average speed in miles per hour (to the nearest tenth) for the first heat of the race?

7. Miles Per Gallon

Goal

To learn how to calculate the familiar rate "miles per gallon" (mpg) in purchasing or caring for a car

Context

One consideration in buying and owning a car is its gas mileage. Because of the cost of gas, cars that can go farther on a gallon of gas cost less to own. Also, mpg fluctuates based on how the car is driven, with highway miles requiring less gas than miles driven in traffic. However, if a car's mpg changes while the type of driving remains the same, it may indicate a problem with the car's mechanical operations.

Teaching Notes

Calculating mpg is a straightforward division problem: miles per gallon = miles driven/gallons of gas burned. Most students use it daily to project costs, diagnose mechanical problems, and recognize the value of more fuel-efficient cars—whether they realize that's what they're doing, or not!

Extension Activities

- Ask students to determine the mpg for their own or their family's car(s). Collect the data and make a classwide chart.

- Make predictions about types and sizes of cars and what kind of gas mileage they might get.

- Have students estimate the miles a car might be driven in a year (the national average is 15,000 miles) and figure out how much it will cost to buy the gas for a given car for a year.

Answers

1. Noemi's car got 29.0 mpg.

2. Tyler's favorite stock car is getting 5.2 mpg.

3. Valli's Land Rover got 24 mpg for the trip.

4. Ryan had driven 337 miles.

5. Mikel should see a mechanic. He got only 16.5 mpg on his trip, well below normal.

6. Caryn and Jemal can go 330 miles on a tank of gas.

7. Miles Per Gallon

Gas mileage is a selling point for many cars. It is a rate of how many miles the car will go on a gallon of gas. This number is different for highway driving and city driving, so it is more accurate to test your own car with your typical driving style. You do not need to burn exactly 1 gallon of gas to calculate miles per gallon (mpg). Just fill your car's tank full, write down your odometer reading, drive around for a few days, and then refill your car (full). Use your odometer to figure out how many miles you drove and the gas receipt to know how many gallons the car used. Divide the number of miles you drove by the number of gallons your car used. This will give you your gas mileage for that period of driving.

Help the following people calculate their gas mileage.

1. Noemi has had her car for a while and knows that it gets pretty good gas mileage. On a recent business trip, she used 16.25 gallons of gas and drove 471 miles. To the nearest tenth of a mile, what was her gas mileage?

2. Tyler loves to watch stock-car racing. He figures that his favorite driver burns 115 gallons of fuel in a 600-mile race. What gas mileage is the stock car getting (to the nearest tenth of a gallon)?

3. Valli has a new Land Rover. On a recent trip, she drove 450 miles and used 18.75 gallons of gas. What was her gas mileage?

(continued)

7. Miles Per Gallon

4. Ryan has a computer in his car that tells him how far he has gone on a trip, how many gallons of gas he has used, and what his gas mileage is. Unfortunately, the device isn't working and after a recent trip he couldn't figure out how far he had traveled. He did know that he had averaged 29.3 miles per gallon and burned 11.5 gallons of gas. How long was the trip?

5. Mikel's car usually gets 28 miles per gallon in mostly highway driving. On a recent long trip mostly on highways, he used 14.2 gallons of gas for 234 miles. Should he see a mechanic to check if his car is working okay?

6. Caryn and Jemal have a new minivan. The van has a 15-gallon tank and gets 22 miles to the gallon. How far can the van go on a tank of gas?

8. Stock-Car Fuel Strategy

Goal

To learn about using information to make mathematical predictions in the setting of a stock-car race

Context

Many students will know about the strategies of racing, but they may not know the ongoing calculations that the crew chief makes to assess when the next pit stop is necessary.

Teaching Notes

Stock-car racing has become one of the most popular spectator sports in the United States. Pit crews and racing teams need to judge the best times to stop in the race to change tires or add fuel. Often they stop while other cars are driving slower yellow-flag laps so that they don't have as much ground to make up when they return to the track.

Extension Activity

Encourage students to watch a stock-car race on TV and listen to the announcers discuss pit-stop strategies. By keeping track of the number of green-flag laps and yellow-flag laps, students should try to calculate the gas mileage the cars will get that day.

Answers

1. No, Scott won't quite make it. He will burn 6.25 gallons in the first

25 miles and 1 gallon in the next 8 miles, leaving him 14.25 gallons in his tank after 33 miles. If the rest of the race is run at full speed, 14.25 gallons will only get Scott 59 miles, but he has 67 left to go.

2. At this point, Scott has used 15 gallons driving 60 miles in the 30 green-flag laps and 3.75 gallons driving 30 miles in the 15 yellow-flag laps. He has used 18.75 gallons and has 3.25 gallons left in his tank.

3. Scott has run 58 miles under the green flag and 12 under yellow, burning a total of 16 gallons. He can go 16 more miles on the 4 gallons of gas left before getting below a level that makes Davey uncomfortable. On a half-mile track, that is 32 more laps.

4. In order to have enough fuel to finish the race, Scott would need to take 28 or more laps under yellow and no more than 122 under green. So far he has completed 10 laps under yellow, so he needs to run 18 more laps under yellow in order to avoid refueling.

5. Scott should expect to refuel at about the 85th lap. (Students who are unsure of how to tackle this algebraically could consider the guess and check method, rounding to whole laps.)

© 2007 Walch Publishing

8. Stock-Car Fuel Strategy

Davey McKenzie is the crew chief for Scott Gaylord Racing, car #00 on the NASCAR West circuit. During a race, one of Davey's most important tasks is to keep track of the fuel level in Scott's car and make sure that Scott doesn't run out of gas. Scott's car has a 22-gallon fuel tank, but Davey doesn't like to let the amount of gas in the tank get below 2 gallons unless the race is about to end. Davey and Scott calculate that they can get about 4 miles per gallon in a race, but they can get 8 miles to the gallon when the race is slowed down by a yellow flag (meaning that laps are run under "caution"). In the following race settings, help Davey decide when to refuel Scott's car.

1. In a 100-lap race at the 1.0-mile Phoenix oval, the first 25 laps are run at full speed, but then there is a tire blowout on another car and the yellow flag comes out for the next 8 laps. If there are no more "caution" laps, can Scott finish the race without refueling?

2. After refueling at the 133-lap mark in a 200-lap race on the 2.0-mile Fontana track, Scott runs 30 laps under a green flag and 15 under yellow in the next 45 laps. How much gas should he have left in his tank at this point?

(continued)

8. Stock-Car Fuel Strategy

3. In a 250-lap race at the $\frac{1}{2}$-mile Mesa Marin Raceway, Scott runs the first 40 laps under green, but in the next 100 laps, 24 laps are under yellow and 76 laps are under green. Assuming that the rest of the laps of the race will run under green, how many more laps should Davey let Scott drive before he needs to come in for refueling? (Remember that Davey wants to keep no fewer then 2 gallons in the gas tank.)

4. In a 150-lap race at the Evergreen Speedway, a .646-mile oval, Scott runs the first 44 laps under green, the next 6 under yellow, the next 9 under green, and the next 4 under yellow. How many more laps would have to be run under yellow in order for Scott to be able to drive the whole race without refueling?

5. For a 150-lap race on a 1.0-mile oval, Scott wants to plan a race strategy. He knows from experience that about 15% of the laps run will be under caution. If Scott wants to refuel after burning 20 gallons of fuel, what lap should he expect to refuel at?

9. Earned Run Average

Goal

To put the concept of rates into a sports context (earned run average) that is familiar to students

Context

The questions in this section were all researched using the *Baseball Encyclopedia.* Most baseball fans love to discuss and manipulate statistics, and ERA is one of the statistics that many fans don't understand.

Teaching Notes

A rate is usually an amount of work accomplished in a given period. It is given as "miles per hour" or "goals per game." Some rates, however, are stated differently. "Dollars per pound" is a rate, but "pound" isn't a period of time. Still other rates don't even use the word "per." In baseball, a pitcher's "earned run average" (ERA) is really a rate. It tells you the rate at which the pitcher gives up earned runs per game. ERA is calculated by dividing the number of innings a pitcher has pitched by 9 to get the number of games pitched. You then take the earned runs the pitcher has yielded and divide by the number of games pitched to learn the ERA. The formula for ERA is as follows:

ERA = Runs/(Innings Pitched ÷ 9)

Extension Activity

The *Baseball Encyclopedia* is an excellent resource for baseball statistics, going back more than a century. Encourage students to search this book, a current newspaper, the Internet, or even get statistics on a school team to prepare several questions like those included in the activity.

Answers

Answers may vary slightly.

1. Gibson's 1968 ERA was 1.12.

2. Seaver's 1974 "worst" ERA was 3.20.

3. Rixey yielded 59 earned runs in 1916.

4. Podres pitched $196\frac{1}{3}$ innings in 1957.

5. Young's career ERA was 2.63.

6. Beck yielded 47 runs in 1934.

7. The Forsch brothers' combined ERA for 1977 was 3.27.

9. Earned Run Average

In baseball, the earned run average (ERA) tells you the rate at which the pitcher gives up earned runs per game. It is calculated using the formula ERA = Runs/(Innings Pitched ÷ 9). Use this formula to answer the following questions.

1. In 1968, Bob Gibson of the St. Louis Cardinals had his greatest season. He pitched $304\frac{2}{3}$ innings and gave up only 38 runs. What was his ERA for the 1968 season?

2. In 1974, Tom Seaver of the New York Mets had his worst season to that point in his career. He pitched 236 innings and gave up 84 runs. What was his ERA for the 1974 season?

3. In 1916, Hall of Famer Eppa Jephtha Rixey had his finest season. He won 22 games for the Phillies and had an ERA of 1.85. Figure out how many earned runs Rixey yielded if he pitched 287 innings that year.

(continued)

9. Earned Run Average

4. Johnny Podres was a star pitcher for the Dodgers. In 1957, he had an ERA of 2.66 while giving up 58 runs. How many innings did he pitch?

5. Denton True "Cy" Young was perhaps the greatest pitcher of all time. In his career, he pitched 7356 innings and gave up 2150 runs. What was his career ERA?

6. Walter "Boom-Boom" Beck had a tough year with the Brooklyn Dodgers in 1934. He had a 7.42 ERA in 57 innings. How many earned runs did he yield?

7. Brothers Ken and Bob Forsch both pitched in the National League in 1977. Bob was 20–7 with the Cardinals, yielding 84 runs in 217 innings. Ken was 5–8 with the Astros, yielding 26 earned runs in 86 innings. What was their combined ERA for the season?

31

10. Currency Trading

Goal

To learn about conversion rates in the context of international currency trading

Context

Most financial reports include "how the dollar is doing against the yen (or deutsche mark)." A "strong dollar" (which can buy more yen) means that Americans can afford to buy more imports from Japan, whereas a "weak dollar" means the Japanese can afford more U.S. exports. International travelers also need to know conversions to get a sense of the value of their purchases.

Teaching Notes

Currencies are traded in the open market every day. The prices are monitored in ratio to one another. As of March 24, 2006, for instance, it took $1.1675 Canadian to purchase $1.0000 U.S. It therefore cost $0.8565 U.S. to purchase $1.0000 Canadian. These ratios could be used to convert with another currency or perhaps to discover how much something costs in one place based on what it costs in another.

Extension Activity

Most financial newspapers have a currency trading report. Ask students to convert between various currencies or determine prices for items in other currencies.

Answers

1. Daver would pay $8510.64 U.S. to buy 1,000,000 yen.

2. Daver's 10,000,000 Mexican pesos can buy 30,039,050 Taiwanese dollars.

3. An Australian dollar will buy 0.5906 of a euro.

4. A U.S. dollar will buy 102.6537 Sri Lankan rupees.

5. Daver's hotel room cost $182.98 in U.S. dollars.

6. The English shirt costs the equivalent of 36.21 euros, so it is a better deal.

7. One Swiss franc bought 4.733 Danish krone.

10. Currency Trading

Andre is a currency trader for National Savings Bank. Each day, he decides how to trade the currency he has for currencies from other countries in an effort to turn a profit. Andre monitors currency prices in ratio to one another. As of March 24, 2006, for example, it took $1.1675 Canadian to purchase $1.00 U.S. Andre determines the number of U.S. dollars needed to purchase one Canadian dollar by dividing $1.00 U.S./$1.1675 Canadian = $0.8565 U.S. To make profitable trades, Andre needs to know the conversion rates between the currencies of many countries at once. He also recently took a business trip to visit some trading partners.

Answer the questions below about currency trading and conversions that Andre has encountered recently.

1. One U.S. dollar buys 117.50 Japanese yen on today's market. How many U.S. dollars would it cost Andre to buy 1,000,000 yen?

2. Andre learns that the Taiwan dollar is worth .3329 Mexican pesos. Andre has 10,000,000 Mexican pesos. How many Taiwanese dollars can he buy?

3. Andre knows that 1 euro will buy 1.6932 Australian dollars. How many euros will 1 Australian dollar buy?

(continued)

10. Currency Trading

4. Sri Lankan rupees carry lesser value than most currency. One Canadian dollar will buy 87.9229 rupees. One Canadian dollar will also buy .8565 U.S. dollars. How many rupees will 1 U.S. dollar buy?

5. The first stop on Andre's trip to Europe was London. His travel agent booked him a hotel room at a rate of 105 pounds per night. How much did Andre pay for the room in U.S. dollars if 1 British pound buys 1.7427 U.S. dollars?

6. Andre's next stop was in Paris. He saw a shirt in a shop window there that was exactly like a shirt he had nearly bought in London. In England, the shirt cost 25 pounds. In Paris, they were asking 38 euros. Andre knew that the euro is worth .6905 pounds. Which shirt was a better deal?

7. Finally, Andre visited Geneva, Switzerland. He had decided not to go to Denmark as originally planned, so he exchanged the 1000 Danish krone he had. The clerk at the currency exchange handed him 211.27 Swiss francs. How many krone did 1 franc buy?

11. Heating Degree Days

Goal

To learn how averages are used in calculating a rate used in weather statistics

Context

Calculating heating degree days originated shortly after World War II; 65° F was picked as the temperature that people would want for their homes. Not only are heating degree days used in the northern winters, but cooling degree days (calculated the same way) are used during the summers. These weather statistics are used to measure how cold (or warm) a given period has been, and they are also central in calculating for heating-oil deliveries.

Teaching Notes

On any given day, the average of the high and low temperatures is calculated. This average is subtracted from 65 to get the heating degree days for that date.

Extension Activities

- Invite a local news station meteorologist to visit your school and do a presentation.

- This type of chart project would be a good opportunity for students to collect their own data and calculate the information.

Answers

Date	High Temp	Low Temp	Degree Days	Total Degree Days
01/07/06	13°	5°	56	56
01/08/06	18°	8°	52	108
01/09/06	23°	12°	47.5	155.5
01/10/06	27°	18°	42.5	198
01/11/06	26°	16°	44	242
01/12/06	32°	20°	39	281
01/13/06	36°	23°	35.5	316.5
01/14/06	28°	22°	40	356.5
01/15/06	27°	20°	41.5	398
01/16/06	15°	10°	52.5	450.5
01/17/06	8°	-2°	62	512.5
01/18/06	12°	0°	59	571.5
01/19/06	19°	5°	53	624.5
01/20/06	23°	11°	48	672.5
01/21/06	27°	13°	45	717.5
01/22/06	26°	17°	43.5	761
01/23/06	33°	20°	38.5	799.5
01/24/06	35°	25°	35	834.5
01/25/06	31°	24°	37.5	872
01/26/06	29°	22°	39.5	911.5
01/27/06	24°	16°	45	956.5

11. Heating Degree Days

Paul Sanchez is a television meteorologist. As well as predicting the coming weather, Paul keeps statistics on past weather. One statistic he keeps is heating degree days. Paul calculates heating degree days by finding the average of the high and low temperature for a given day and then subtracting it from 65°. Paul uses this number to show people how cold it has been and also to help heating-oil companies plan their deliveries. Paul has kept the following table of high and low temperatures for three weeks in January. Can you help him finish calculating the number of heating degree days each day and the running total for the month?

Date	High Temp	Low Temp	Degree Days	Total Degree Days
01/07/06	13°	5°		
01/08/06	18°	8°		
01/09/06	23°	12°		
01/10/06	27°	18°		
01/11/06	26°	16°		
01/12/06	32°	20°		
01/13/06	36°	23°		
01/14/06	28°	22°		
01/15/06	27°	20°		
01/16/06	15°	10°		
01/17/06	8°	-2°		
01/18/06	12°	0°		
01/19/06	19°	5°		
01/20/06	23°	11°		
01/21/06	27°	13°		
01/22/06	26°	17°		
01/23/06	33°	20°		
01/24/06	35°	25°		
01/25/06	31°	24°		
01/26/06	29°	22°		
01/27/06	24°	16°		

12. Batting Average

Goal

To put the concept of rates into a sports context (batting average) that is familiar to students

Context

The questions in this section were all researched using the *Baseball Encyclopedia.* Most baseball fans love to discuss and manipulate statistics, and batting average is a popular one.

Teaching Notes

A rate is usually an amount of work accomplished in a given period. Some rates, however, are not stated "_____ per _____." For example, in baseball, a player's batting average is really a rate. It tells you the rate of hits a batter gets per official at-bat. The batting average is calculated by taking the number of hits a batter gets and dividing it by the number of official at-bats (opportunities) he or she has. When a batter is walked, hit by a pitch, or sacrifices (bunts or flies out to the team's benefit), it does not count as an official at-bat. The formula for batting average is hits/official at-bats.

Extension Activities

The *Baseball Encyclopedia* is an excellent resource for baseball statistics going back more than a century. Encourage students to search this book, a current newspaper, the Internet, or even get statistics on a school team to prepare several questions like those included in the activity.

Answers

1. Duffy's 1894 batting average was .440.

2. Dropo's career batting average was .270.

3. Ruth had 42 hits in World Series play.

4. Ferguson had 91 hits in 1878.

5. Versalles had 667 at-bats in 1965.

6. Gwynn had 419 at-bats in 1994.

7. Cobb batted .240 in his only bad year.

8. The DiMaggio brothers' combined batting average for 1940 was .318.

9. Robinson had 589 at-bats in his first season in the majors.

12. Batting Average

In baseball, a player's batting average is the rate of hits per official at-bats. Official at-bats exclude when a batter is walked, hit by a pitch, or sacrifices (bunts or flies out to their team's benefit). To calculate a player's batting average, divide the player's hits by his or her official at-bats. Answer the following questions about batting averages.

1. In 1894, Hugh Duffy, a center fielder for the Boston Nationals, had the best season for batting average in history. He led the league with 237 hits in 539 official at-bats. What was his batting average in this historic season?

2. Walt "Moose" Dropo had a 13-year major league career with the Red Sox, Tigers, White Sox, Reds, and Orioles. In his career, he batted 4124 times and recorded 1113 hits. What was Dropo's career batting average?

3. George Herman "Babe" Ruth was arguably the greatest hitter of all time. He played in 10 World Series with the Red Sox and Yankees, batting .326 in 129 at-bats in the series. How many hits did Ruth have in the World Series?

4. Robert Vavasour "Death to Flying Things" Ferguson had his best batting year with the Chicago Nationals in 1878. He had a batting average of .351 in 259 at-bats. How many hits did he have?

(continued)

12. Batting Average

5. In 1965, Zoilo Versalles of the Minnesota Twins led the league in official at-bats. He had a batting average of .273, knocking out 182 hits. How many official at-bats did he have?

6. Tony Gwynn is considered one of the greatest hitters of the modern era. During the strike-shortened 1994 season, he had his highest batting average ever, accumulating 165 hits and a .394 batting average. How many at-bats did Gwynn have in this season?

7. Tyrus Raymond "Ty" Cobb has the highest career batting average ever. In his 24-year career, he batted below .300 only once! It was his first year in the majors and he had 36 hits in 150 at-bats. What was Cobb's career-low batting average in 1905?

8. Dom, Joe, and Vince DiMaggio were undoubtedly the greatest set of three brothers to play in the majors. In 1940, Dom had 126 hits in 418 at-bats, Joe had 179 hits in 508 at-bats, and Vince knocked out 104 hits in 360 at-bats. What was the DiMaggio family batting average in 1940?

9. In 1947, Jackie Robinson broke the color barrier in the major leagues. Despite the pressures of the season, Robinson rapped out 175 hits and batted .297. How many at-bats did he have in 1947?

13. Slugging Average

Goal

To put the concept of rates into a sports context (slugging average) that is familiar to students

Context

The questions in this section were all researched using the *Baseball Encyclopedia.* Most baseball fans love to discuss and manipulate statistics, and slugging average is sometimes a confusing one.

Teaching Notes

In baseball, a player's "slugging average" is really a rate. It tells you the number of bases a batter earns per official at-bats. An official at-bat is recorded whenever a batter gets up and does not get a walk, get hit by a pitch, or hit into a sacrifice (intentional) out. The formula for slugging average is

$$\frac{\text{No. of singles} + 2 \times (\text{No. of doubles}) + 3 \times (\text{No. of triples}) + 4 \times (\text{No. of home runs})}{\text{Number of official at-bats}}$$

Extension Activity

The *Baseball Encyclopedia* is an excellent resource for baseball statistics going back more than a century. Encourage students to search this book, a current newspaper, the Internet, or even get statistics on a school team to prepare questions like those included in the activity.

Answers

1. Ruth's best-ever slugging average was .847.

2. Rodriguez had a .408 slugging average in 1995.

3. Maranville's 1919 slugging average was .377.

4. Mellilo's 1931 slugging average was .407.

5. Parker has a career World Series slugging average of .415.

6. Gehrig's career slugging average was .632.

7. All of Reilley's 13 hits were singles.

8. Meixell's hit was a single.

13. Slugging Average

In baseball, slugging average is calculated by taking the number of bases that a batter achieves through hitting (singles count as one, doubles as two, and so forth), and dividing by the number of official at-bats (opportunities) he or she has. When a batter is walked, hit by a pitch, or sacrifices (bunts or flies out to the team's benefit), it does not count as part of either total. The formula for slugging average is:

$$\frac{\text{No. of singles} + 2 \times (\text{No. of doubles}) + 3 \times (\text{No. of triples}) + 4 \times (\text{No. of home runs})}{\text{Number of official at-bats}}$$

Complete the slugging average data by answering the questions below.

1. Babe Ruth owns 6 of the top-10 all-time highest slugging averages for a year. In 1920, he set the record for highest slugging average for a season ever when he hit 54 home runs, 9 triples, 36 doubles, and 73 singles in his 458 at-bats. What was his slugging average in this historic season?

2. Alex Rodriguez, now the star third baseman for the New York Yankees, was a shortstop for the Seattle Mariners in 1995. In 142 at-bats that year, he had 20 singles, 6 doubles, 2 triples, and 5 home runs. What was his 1995 slugging average?

3. Walter James Vincent "Rabbit" Maranville was a star for the Boston Braves. In 1919, he knocked out 95 singles, 18 doubles, 10 triples, and 5 home runs. If he had 480 official at-bats, what was his slugging average?

(continued)

13. Slugging Average

4. Oscar Donald "Spinach" Mellilo played for the 1931 St. Louis Browns. In 617 at-bats that year, he hit 142 singles while bashing 34 doubles, 11 triples, and 2 home runs. What was his slugging average for the 1931 season?

5. David Gene "The Cobra" Parker went to the World Series three times, in 1979 with the Pittsburgh Pirates, and again with the Oakland Athletics in 1988 and 1989. In his World Series career, Parker batted 53 times, getting 10 singles, 4 doubles, and 1 home run. What was his World Series slugging average?

6. Lou Gehrig held the record for consecutive games played until it was broken in 1996 by Cal Ripken, Jr. Before his career and life were cut short by ALS (amyotrophic lateral sclerosis), Gehrig had 8001 at-bats, hitting 493 home runs, 162 triples, 535 doubles, and 1531 singles. He finished third on the all-time career list with what slugging average?

7. Alexander Aloysius "Duke" Reilley wasn't very big (his teammates called him "Midget"). He played 20 games in 1909 for Cleveland. In his 62 at-bats, he had 13 hits. His slugging average was .210. How many singles did Reilley hit?

8. Merton Merrill "Moxie" Meixell played only 2 games in the major leagues, both with the 1912 Cleveland Spiders. He had 2 at-bats and 1 hit. His career slugging average was .500. What type of hit did he get?

14. Exercise and Calories

Goal

To learn about rates in the context of exercise and burning calories

Context

Our society is very conscious of exercise for health and fitness. The body keeps its energy up by burning calories. When a person runs on a flat surface, he or she burns calories at a relatively constant rate. Running faster shortens the workout, but running 5 miles in 30 minutes doesn't burn more calories than running 5 miles in an hour—it just burns them in less time.

Teaching Notes

Two factors go into calculating the number of calories a runner burns while running on a flat surface: the weight of the runner and the speed and length of time he or she is running. The formulas for these calculations are as follows:

$$K = {}^{13}/_8 \, W - {}^5/_4$$

$$C = KD$$

K is the number of calories burned per mile of running, W is the weight of the runner (in kilograms), C is the number of calories burned, and D is the number of miles run.

Extension Activities

- Many physiology texts include information about energy expenditure and calories. Students could research other forms of exercise and the numbers involved in those forms.

- Many exercise machines have computers that calculate the calories burned in a period of exercise. Sample numbers from these machines could be used to build equations for that form of exercise.

Answers

1. Samír will burn 119 calories per mile.

2. Tyler will burn 400 calories in his run.

3. Emily ran 6.87 miles.

4. Unice will burn 93 calories per mile.

5. Terrence would burn 289 calories in her run.

6. Natasha ran 8.7 miles.

7. Chris weighs 70 kilograms.

14. Exercise and Calories

Talesha is a personal trainer. She assists individuals in setting up healthy exercise patterns and sticking to them. Good health is a combination of sufficient and safe exercise and good nutrition. For their exercise programs, Talesha advises many of her clients to run on a flat surface. She makes recommendations about how fast and how far they should run for them to exercise safely. Talesha also determines the number of calories that a client will burn during their run.

The formulas for these calculations are $K = {}^{13}\!/_8\, W - {}^5\!/_4$ and $C = KD$ where K is the number of calories burned per mile of running, W is the weight of the runner (in kilograms), C is the number of calories burned, and D is the number of miles run. Help Talesha calculate the data she needs for her clients by answering the questions below. (Round answers to the nearest tenth.)

1. Samír weighs 74 kilograms. How many calories per mile will he burn running on a flat surface?

2. Tyler weighs 50 kilograms. How many calories will he burn in a 5-mile run?

3. Emily weighs 52 kilograms. She burned 572 calories in her run yesterday. How far did she run?

4. Unice weighs 58 kilograms. How many calories per mile will she burn running on a flat surface?

5. Terrence weighs 60 kilograms. How many calories would he burn in a 3-mile run?

6. Natasha weighs 58 kilograms. She burned 810 calories in her run yesterday. How far did she run?

7. Chris ran $9\,{}^1\!/_3$ miles. Talesha calculated that he burned 1050 calories. How much does Chris weigh?

44

15. Tipping

Goal

To learn about calculating tips or gratuities in restaurants, from the server's point of view

Context

Almost everyone eats in a restaurant at some point, and we usually leave a tip for the server. Customers usually leave 15–20% for a tip.

Teaching Notes

Rates are often given in percentages. To calculate a tip, you simply multiply the food bill (usually before taxes are added) by the percentage of tip you want to leave: food bill × percent = tip. Of course, this formula can be manipulated to calculate the size of the bill or the percent that the tip represents of that bill. You may want to discuss the wage structure for servers (below minimum wage base + tips) so that students understand the importance of tips. An explanation of the reason for a restaurant's policy about large parties might also be helpful.

Extension Activities

Students can interview restaurateurs and service staff at local restaurants to learn more about tips. They can also brainstorm lists of other professions where employees receive tips.

Answers

1. Dalit added a $17.03 gratuity to the bill.

2. The anniversary couple's tip represented 17.5% of the food bill.

3. The businesswoman left a $3.11 tip.

4. The man's tip represented 10.8% of his food bill.

5. Dalit charged the birthday party a $33.52 gratuity.

15. Tipping

Dalit has a great job as a server at a restaurant. She takes orders, serves food, and ensures a pleasant dining experience for patrons. According to restaurant policy, if there are 6 or more people in a party, Dalit has to add on a 15% gratuity (tip) to the bill before taxes. If a table has 5 or fewer patrons, they may choose how much they would like to tip. The standard tip used to be 15%. However, many people now tip between 20% and 25%, depending on the quality of the service they receive. Tips are very important to servers because their hourly wage is about half the minimum wage. Fortunately, Dalit is good at her job!

To calculate a tip, you simply multiply the sum of the food and beverage bill (before sales tax is added) by the percentage of the tip you want to leave: food sum × percent = tip.

Answer the following questions about Dalit's tip income from last Friday night.

1. Dalit's first table was a family of 7. Everyone had sodas, salads, and entrees. Their bill came to $113.52. Dalit had to calculate the gratuity to include in the bill. How much gratuity did she include?

2. Dalit also served a couple celebrating their anniversary. They had a bottle of champagne, salads, entrees, and dessert. Their bill was for $85.63 before tax, and they tipped $15.00. What percent of their food bill did the tip represent?

3. Dalit served a businesswoman, dining alone, who just wanted soup, salad, and water. Her bill came to $12.43. She appreciated that Dalit was friendly and efficient and wanted to tip 25%. How much of a tip did she leave?

4. A fussy customer came in and ordered several items that were not on the menu. Unfortunately, a busperson also spilled the man's water while pouring it for him. The diner felt that he didn't get good service, and he left only a $2.00 tip on a bill of $18.50. What percent tip was this?

5. Four couples came in together to celebrate someone's birthday. Their bill came to $223.45. How much gratuity did Dalit charge them?

16. Produce, Meats, and Cheeses

Goal

To learn how to calculate with rates, using produce and deli prices in a grocery store

Context

Grocery stores are valuable sources of math problems. Nearly every item in the produce, deli, and meat sections of a grocery store has its price given by the pound. As students learn to shop and estimate, they will begin to see differences in prices and become better mathematicians and more careful shoppers, too.

Teaching Notes

Perhaps the most common rate not related to speed or work is price per pound. Many everyday items are purchased by the pound. To determine the price of an item, you weigh it and then multiply the weight in pounds by the price per pound. The formula is as follows:

price = no. of pounds × price per pound

Extension Activity

Grocery stores are usually willing to have small groups visit on field trips. Students can learn about pricing of produce, meats, and cheeses, and how scales are used in pricing items.

Answers

Item	Weight	Price per pound	Cost
apples	2.2 pounds	$1.89	$4.16
American cheese	9 ounces	$3.29	$1.85
bananas	2.45 pounds	$0.89	$2.18
bologna	1/2 pound	$3.29	$1.65
capicola	1/4 pound	$6.39	$1.60
cocktail shrimp	1.4 pounds	$6.89	$9.65
green beans	1.2 pounds	$1.59	$1.91
mint (fresh)	1.5 ounces	$2.09	$0.20
strip steak	1.7 pounds	$4.29	$7.29
oranges	5.2 pounds	$1.49	$7.75
papaya	10 ounces	$2.79	$1.74
potato salad	8 ounces	$1.99	$1.00
potatoes	10.4 pounds	$0.59	$6.14
red peppers	.75 pounds	$3.99	$2.99
salad bar	10 ounces	$4.59	$2.89
snow peas	5 ounces	$2.39	$0.74
Swiss chard	.4 pounds	$0.89	$0.36
tomatoes	1.3 pounds	$3.19	$4.15
turkey breast	3/4 pound	$5.09	$3.82
watermelon	7.6 pounds	$0.69	$5.24

Name _____   Date _____

16. Produce, Meats, and Cheeses

Joelle is a careful shopper. She keeps track of how much she spends and whether an item is within her price range. Joelle likes fresh produce, meats, and cheeses, all of which are sold at a given price per pound. On a recent trip to the store, Joelle came across the following prices and weights. Calculate how much Joelle spent on each item below. Remember, there are 16 ounces in a pound.

Item	Weight	Price per pound	Cost
apples	2.2 pounds	$1.89	
American cheese	9 ounces	$3.29	
bananas	2.45 pounds	$0.89	
bologna	$\frac{1}{2}$ pound	$3.29	
capicola	$\frac{1}{4}$ pound	$6.39	
cocktail shrimp	1.4 pounds	$6.89	
green beans	1.2 pounds	$1.59	
mint (fresh)	1.5 ounces	$2.09	
strip steak	1.7 pounds	$4.29	
oranges	5.2 pounds	$1.49	
papaya	10 ounces	$2.79	
potato salad	8 ounces	$1.99	
potatoes	10.4 pounds	$0.59	
red peppers	.75 pounds	$3.99	
salad bar	10 ounces	$4.59	
snow peas	5 ounces	$2.39	
Swiss chard	.4 pounds	$0.89	
tomatoes	1.3 pounds	$3.19	
turkey breast	$\frac{3}{4}$ pound	$5.09	
watermelon	7.6 pounds	$0.69	

17. Unit Pricing

Goal

To use rates in the familiar concept of unit pricing of groceries

Context

Most of us shop in grocery stores at least periodically. So many people comparison shop that stores advertise their unit prices. As part of budgeting the family money, students should know how to comparison shop.

Teaching Notes

A rate is usually an amount of work or accomplishment achieved in a given period. It is generally given as "_____ per _____." It can also be a value for a certain amount of time or money. We can talk about "miles per hour," "goals per game," "dollars per hour," or "dollars per pound." In the case of unit pricing, the idea is to get a comparative price between two items of different size and price. The formula for calculating price per ounce is as follows:

cost of package/size of package
(in ounces) = cost per ounce

Extension Activity

Your local grocery store is a wonderful resource for learning unit pricing. The management may be willing to give you a tour and discuss their procedures, or you can suggest that students go to the store and compare the prices on some products. Have them ask why larger containers tend to cost less per unit.

Answers

1. The store brand shampoo is cheaper at 13.5 cents per ounce.

2. The loose carrots at 35 cents per pound are cheaper.

3. The 10-pound sack at 12.9 cents per pound is the best deal.

4. The 20-pound bushels are cheapest at 84 cents per pound.

5. The 42-ounce bottle is cheaper at 17.8 cents per ounce.

6. The frozen lemonade is cheapest at $1.17 per gallon, or $.92 per ounce.

7. Debba should buy six 8-ounce bags, costing $9.54.

8. Breda should buy the 64-ounce can for 23.7 cents per ounce.

17. Unit Pricing

Most of us are very careful when we go to the grocery store. We want to save money and still buy what we want. Because of this, most grocery stores now have unit pricing on their shelves, so you can compare costs. For instance, a box of Cleano Soap might sell for $1.45, and a box of Scrubbie Soap might cost $2.15. However, if the Cleano Soap box contains 12 ounces, and the Scrubbie Soap box contains 20 ounces, Scrubbie Soap may be a better buy. If you calculate the price per ounce, you may find that it is more economical to buy the more expensive, but bigger, box.

Help the following careful shoppers choose the best value.

1. Drew needs some new shampoo. He notices that a store brand costs $3.24 for a 24-ounce bottle, and a name brand costs $2.89 for a 19-ounce bottle. If Drew wants to spend as little as possible, which brand should he buy?

2. Noma is deciding whether she should buy her carrots loose or in a 3-pound bag. The loose carrots cost 35 cents per pound and a 3-pound bag of carrots is on sale for $1.10. Which should she buy?

3. Marcus is buying flour to bake bread. He can buy flour in a 5-pound sack, a 10-pound sack, or a 20-pound sack. The 5-pound sack is $.70, the 10-pound sack is $1.29, and the 20-pound sack is $2.75. Which sack represents the best deal?

(continued)

17. Unit Pricing

4. Betsy loves to can tomatoes. At the corner stand, she can buy a 2-pound package for $1.98, a 5-pound package for $4.75, and a 20-pound bushel for $16.80. Assuming that she wants 40 pounds of tomatoes, what size packages should she buy?

5. Mr. Neilan thinks olive oil is good for his health. He is willing to buy big jars, because it doesn't go bad on his shelf. Golden Olive has a 17-ounce bottle for $3.29 and a 42-ounce bottle for $7.48. Which bottle should he buy to pay the least amount per ounce?

6. Lowell likes lemonade. He can buy a 1-gallon (128-ounce) jug of lemonade for $1.29. He can also buy frozen lemonade concentrate for $0.55, which makes 60 ounces, or powdered lemonade that makes 2 gallons (256 ounces) for $2.89. Lowell knows that he will finish whatever size container he buys. Which container should he buy?

7. Debba is making pizzas. She needs to choose among three bags of shredded mozzarella. One contains 8 ounces and costs $1.59. One contains 12 ounces and costs $2.49. One contains 16 ounces and costs $3.29. If Debba needs 48 ounces of cheese for her pizzas, how many of which type of bag should she buy?

8. Breda is buying wallpaper remover for her bedroom. She learns that the hardware store carries three sizes of cans: 15 ounces for $4.29, 33 ounces for $8.16, and 64 ounces for $15.19. She wouldn't mind having some left over if she doesn't use it all, so which size can should she buy?

18. Rate of Pay

Goal

To get a sense of calculating with time and rates of pay

Context

Time cards are still prevalent, especially in the types of jobs that students hold. Not only do they need to know how to deal with their own cards, but someday they may be in charge and need to complete others' cards.

Teaching Notes

The difficulties here lie in calculating how much time a person has worked each day, adding those numbers, and then multiplying by a decimal rate. Some students will assume that they can use their calculator to find hours worked, and for a shift from 9:05 A.M. to 3:35 P.M. will put the operation "9.05 − 3.35 =" on their calculator. You may want to have them figure out why this doesn't work, and develop a strategy for what might. You also may want to discuss the implications of rounding and how to ensure fairness.

Extension Activity

If students have hourly jobs, have them work out their own earnings.

Answers

Answers may vary slightly.

1. Carlos worked 7.5 + 7.5 + 7.75 + 7.5 + 7.633 = 37.88 hours to earn $333.34.

2. Anthony worked 7.333 + 7.667 + 7.417 + 7.8 + 7.483 = 37.70 hours to earn $254.48.

3. Mustafa worked 7.583 + 7.583 + 7.533 + 7.833 + 7.783 = 38.32 hours to earn $319.21.

4. Robert worked 7.067 + 6.9 + 6.95 + 7.117 + 7.083 = 35.12 hours to earn $345.93.

5. Judy worked 4.083 + 4.15 + 4.333 + 7.0 + 4.033 = 23.60 hours to earn $244.26

6. Amani worked 4.167 + 4.0 + 4.167 + 4.167 + 4.0 = 20.50 hours to earn $135.30.

7. Gail worked 7.0 + 7.583 + 5.917 + 5.5 + 6.917 = 32.92 hours to earn $534.95.

18. Rate of Pay

Traci is responsible for her company's payroll. She gets time cards from seven employees each week, calculates how many hours they have worked, and then calculates how much they are to be paid, based on their rate of pay.

Calculate how much each of the following employees should be paid.

1. Name: Carlos Monteagudo

Day	Time in	Time out	Hours worked
Monday	6:30 A.M.	2:00 P.M.	_____
Tuesday	6:35 A.M.	2:05 P.M.	_____
Wednesday	6:30 A.M.	2:15 P.M.	_____
Thursday	6:32 A.M.	2:02 P.M.	_____
Friday	6:55 A.M.	2:33 P.M.	_____

Total hours worked: _____

Rate of pay per hour: $8.80

Payment due:_____

2. Name: Anthony Moor

Day	Time in	Time out	Hours worked
Monday	7:00 A.M.	2:20 P.M.	_____
Tuesday	6:55 A.M.	2:35 P.M.	_____
Wednesday	7:03 A.M.	2:28 P.M.	_____
Thursday	6:45 A.M.	2:33 P.M.	_____
Friday	6:58 A.M.	2:27 P.M.	_____

Total hours worked: _____

Rate of pay per hour: $6.75

Payment due: _____

(continued)

18. Rate of Pay

3.   Name: Mustafa Khader

Day	Time in	Time out	Hours worked
Monday	6:55 A.M.	2:30 P.M.	_____
Tuesday	6:55 A.M.	2:30 P.M.	_____
Wednesday	6:58 A.M.	2:30 P.M.	_____
Thursday	6:45 A.M.	2:35 P.M.	_____
Friday	6:58 A.M.	2:45 P.M.	_____

Total hours worked: _____

Rate of pay per hour: $8.33

Payment due: _____

4.   Name: Robert Worley

Day	Time in	Time out	Hours worked
Monday	10:06 A.M.	5:10 P.M.	_____
Tuesday	10:08 A.M.	5:02 P.M.	_____
Wednesday	10:10 A.M.	5:07 P.M.	_____
Thursday	10:01 A.M.	5:08 P.M.	_____
Friday	11:00 A.M.	6:05 P.M.	_____

Total hours worked: _____

Rate of pay per hour: $9.85

Payment due: _____

5.   Name: Judy Zartman

Day	Time in	Time out	Hours worked
Monday	9:00 A.M.	1:05 P.M.	_____
Tuesday	9:05 A.M.	1:14 P.M.	_____
Wednesday	8:55 A.M.	1:15 P.M.	_____
Thursday	6:00 A.M.	1:00 P.M.	_____
Friday	9:02 A.M.	1:04 P.M.	_____

Total hours worked: _____

Rate of pay per hour: $10.35

Payment due: _____

(continued)

18. Rate of Pay

6. Name: Amani Boling-Khader

Day	Time in	Time out	Hours worked
Monday	2:00 P.M.	6:10 P.M.	_____
Tuesday	2:05 P.M.	6:05 P.M.	_____
Wednesday	2:08 P.M.	6:18 P.M.	_____
Thursday	2:03 P.M.	6:13 P.M.	_____
Friday	2:07 P.M.	6:07 P.M.	_____
Total hours worked: _____			
Rate of pay per hour: $6.60			
Payment due: _____			

7. Name: Gail Boling

Day	Time in	Time out	Hours worked
Monday	7:30 A.M.	2:30 P.M.	_____
Tuesday	10:00 A.M.	5:35 P.M.	_____
Wednesday	10:50 A.M.	4:45 P.M.	_____
Thursday	6:45 A.M.	12:15 P.M.	_____
Friday	7:25 A.M.	2:20 P.M.	_____
Total hours worked: _____			
Rate of pay per hour: $16.25			
Payment due: _____			

19. Population Growth Rates

Goal

To learn about growth rates and population while practicing exponent use

Context

Population growth is a global issue because of concerns about whether the earth can sustain a larger population. On a more local scale, population growth and decline has implications for tax structures and representation decisions with respect to local and national government.

Teaching Notes

The formula for population growth is fairly easy to derive. Given the rate of growth, r, and the current population (at time zero), P_o, the population after 1 year is given by the following formulas:

$$P(1) = P_o + P_o(r) = P_o(1 + r)$$

$$P(2) = P_o(1 + r) + P_o(1 + r)(r) = P_o(1 + r)^2$$

Therefore the population at a given time (t) is determined by this formula:

$$P(t) = P_o(1 + r)^t$$

This formula is fairly easy for students to manipulate if they know exponents.

Extension Activity

Using online resources, almanacs, or statistics compendia, students can research, graph, and discover the growth rate of a population. They can then project future population and the political-economic implications of their projections.

Answers

Answers may vary slightly.

1. Tyler's population will be 155,085 in 2009 at this rate of growth.

2. St. Johnsbury's population will be 7868 in 2012 at this rate of growth.

3. Evanston's population will be 84,983 in 2020 at this rate of growth.

4. Norman's population will be 163,459 in 2010 at this rate of growth.

5. Truth or Consequence's population will be 7649 in 2010 at this rate of growth.

6. Philadelphia's population will be 888,518 in 2020 at this rate of decrease.

19. Population Growth Rates

Anna works as a statistician for the U.S. Census Bureau. She plots and calculates population changes to determine growth in certain areas and then attempts to project future growth. Her projections are used in setting tax policies and in calculating how many representatives a state or local district may elect. Anna uses exponents to help her quickly complete her calculations. The formula, with the rate of growth, r, and the current population (at time zero), P_0, gives the population after one year:

$$P(1) = P_0 + P_0 (r) = P_0 (1 + r)$$

$$P(2) = P(1) + P(1) (r) = P_0 (1 + r) + P_0 (1 + r) (r) = P_0 (1 + r) (1 + r) = P_0(1 + r)^2$$

Therefore the population at any given time (t) is determined by the formula $P(t) = P_0(1 + r)^t$.

Answer the following questions that Anna has encountered recently.

1. Tyler, Texas, had a population of 83,650 in 2000. They have experienced significant growth of 7.1% per year recently. Anna has been asked to project Tyler's population in 2009. What estimate should she give based on past growth?

2. The population of St. Johnsbury, Vermont, grows at a slow but fairly constant rate of 0.5% per year. In 2004, the population was 7560. What will the population be in 2012?

3. Evanston, Illinois, has experienced a slow growth rate of 0.8% per year for the last several years. Using this as a model and the knowledge that Evanston's population was 74,811 in 2004, what would you expect Evanston's population to be in 2020?

(continued)

57

19. Population Growth Rates

4. In 2000, Norman, Oklahoma, had a population of 95,694. If Norman has been experiencing an active growth rate of 5.5% per year, what population would you project for the year 2010?

5. Truth or Consequences, New Mexico, had a population of 7163 in 2004. The population there has grown at a constant rate of 1.1%. Approximately how many inhabitants will there be in 2010?

6. The city of Philadelphia had 1,470,571 residents in 2004. Since then, Philadelphia has seen a decrease in population of about 3.1% per year. At this rate, how many Philadelphians will there be in 2020?

20. Half-Life

Goal

To learn about decay rates and half-lives in the context of chemistry while practicing exponent use

Context

Chemists and physicists figure out decay rates and half-lives of concentrations of chemicals to determine where they are in their life cycles. Therefore, environmentalists are very concerned with the half-lives of toxins and pollutants, and the rates at which they will decay into (or away from) being dangerous. Similarly, archaeologists often use the carbon-14 content of an object to determine approximately when it was alive or formed.

Teaching Notes

The formula for half-life is $A(t) = A_o(2)^{-t/h}$, where A is the amount or concentration of something, A_o is its original concentration or amount, t is the time since it came into being, and h is its half-life. Because solving for t or h requires logarithms (unless they are obvious), these problems use $A(t)$ and A_o as the unknowns.

Extension Activities

A local museum curator or environmentalist would probably be willing to discuss how they use decay rates. Also, your school's physics or chemistry teacher would probably love to get the chance to address students about this scientific approach in a math classroom.

Answers

1. Selma would expect to find a concentration of 3.04% if Professor Perez is right.

2. When the leak occurred, the toxin's concentration was 42.3%.

3. Selma would expect to find a concentration of 1.15% if Tad is right.

4. Yes, it is down to 6.2 parts per million.

5. Yes, if the body was at 1 percent (.01) concentration in 1935, it would be expected to have 8.01×10^{-6} concentration now. The concentration of 2.0×10^{-5} is clearly more than 8.01×10^{-6}, which means the concentration was more than 1 percent at the time of death in 1935.

© 2007 Walch Publishing

20. Half-Life

Chemicals and chemical compounds decay steadily over time. Selma's job involves measuring this decay. She receives samples of various materials from around the world and makes predictions based on her knowledge of half-lives and her chemical analyses. Selma uses the following formula for half-lives: $A(t) = A_o(2)^{-t/h}$, where A is the amount or concentration of something, A_o is its original concentration or amount, t is the time since it came into being, and h is its half-life. Help Selma by answering the questions below.

1. Professor Perez has been running an archaeological dig in the Middle East for several years. Some of the dating of the site is tricky. His excavators recently dug up a charcoal fire and he sent Selma a sample. He expects that the charcoal dates from 2100 B.C.E. Selma knows that the concentration of carbon-14 in charcoal when it is formed is 5% and that the half-life of carbon-14 is 5730 years. If Professor Perez is right about his dates, about what concentration of carbon-14 might Selma expect to find in 2007?

2. At a superdump site, a toxin has been leaking for 20 days when an inspector discovers it. Selma knows that the toxin has a half-life of 6 days. The inspector reports that the concentration of the toxin upon discovery is still 4.2% in the area. What was the concentration of the toxin when the leak occurred?

3. Tad Baker runs an archaeological dig at a colonial fort. Most of the artifacts that have been found date from after 1725. One day an excavator discovers an old foundation made of oak, which appears to be older. Tad thinks it may date back to 1675. He sends Selma a sample. She knows that the half-life of carbon-14 is 5730 years and that oak has an initial concentration of 1.2%. What concentration of carbon-14 would Selma expect to find in 2007 if Tad is right about the age of the wood?

(continued)

20. Half-Life

4. After a recent oil spill in Portland Harbor, the Environmental Protection Agency (EPA) found that the concentration of oil content in the water was decaying with a half-life of 30 days. The concentration after the spill was 5 parts per thousand. The EPA knows that the water will be safe for all ecological purposes when the concentration reaches 7 parts per million. It has been 290 days since the spill. Is the water safe yet?

5. In 1998, in an effort to solve a murder committed in 1935, the police exhume the body of the victim. They want to see if there is a trace of a poison in the skeleton. They know that the poison decays with a half-life of 7 years. To be fatal, this poison must reach a 1% concentration throughout the body. They discover that the poison concentration in the bones is only 2.0×10^{-5} in 2007. Could this poison have killed the victim in 1935?

21. Mortgage Rates

Goal

To learn about interest rates and mortgages, also learning the net present value of an annuity formula

Context

Mortgage rates fluctuate slowly over the years. When people buy a house or make another major purchase, they often do so by borrowing from the bank and agreeing to make monthly payments on the loan. It is always a shock to discover the high cost of this kind of long-term credit.

Teaching Notes

The net present value formula is difficult to derive before pre-calculus, but students can manipulate the formula, as long as they know how to use exponents. (Use of scientific calculators is recommended for calculating exponents.) The formula is as follows:

$$A = R \, (1 - (1 + i)^n \,) \,/\, i$$

A is the present value of the loan (purchase price of the house), R is the monthly payments, i is the monthly interest rate (annual rate divided by 12), and n is the number of months in the loan period (for a 30-year mortgage, $n = 360$).

Extension Activity

Mortgage rates are listed online and in local newspapers. Students can research them and local house prices. Students might also call local banks and loan companies to learn about rates and qualifying for loans. Bankers are often willing to come to the classroom to discuss these and other products. Many students will be taking out student loans in the near future, and they will see that student loan repayment is similar to a mortgage loan.

Answers

1. a. The couple will have to pay $880.52 each month.

   b. Their cost would be a whopping $316,987.20.

2. The family's payment will be $1125.59 each month.

3. The newlyweds can afford a loan of $139,763.97.

4. payments on a 15-year mortgage: $1902.56; payments on a 30-year mortgage: $1540.29

21. Mortgage Rates

Erica Burns is a mortgage broker. She helps customers determine how much they can afford to spend on a home and what loan program best meets their needs. Erica uses the net present value formula to determine principal and interest payments on her clients' home mortgages. The formula is $A = R (1 - (1 + i)^n) / i$, where A is the present value of the loan (purchase price of the house), R is the monthly payments, i is the monthly interest rate (annual rate divided by 12), and n is the number of months in the loan period (for a 30-year mortgage, $n = 360$).

For example, on a 20-year $75,000 loan at 6.75% interest, the monthly payments would be calculated as follows: $75,000 = R (1 - (1 + (.0675 / 12)^{240}) / (.0675 / 12) = 131.5159565$. This amount is divided into 75,000 to determine $R = \$570.27$.

Help Erica answer the following questions for her clients.

1. a. A couple has chosen a house that they want to buy, and they need to borrow $120,000 to make the purchase. Erica finds them a 30-year mortgage at 8%. What will their monthly mortgage payments be?

    b. How much will the couple end up paying for their house if they make 360 payments of the amount you found?

2. A family of five needs a bigger house. They ask Erica to find them a 30-year mortgage for $165,000 that is affordable. She finds a 7.25% mortgage rate for families. How much will their monthly payments be?

3. A pair of newlyweds calculate that they can afford $1050 per month in mortgage payments. Erica finds a first-time-buyer loan for 30 years at 8.25%. How much can they afford to borrow to buy their house?

4. A young family has recently increased their income and wants to move into a bigger home. They like a house for which they would have to borrow $215,000. Erica offers them a 15-year mortgage at 6.75% or a 30-year mortgage at 7.75%. What would their monthly payments be on each loan?

22. Auto Loan Rates

Goal

To learn about interest rates and automobile loans, also learning the net present value of an annuity formula

Context

Many high-school students are getting their licenses and have first-time use of a car, either their own or a family member's. Most people who are new to cars think of costs as being the gas and perhaps the insurance. When they are buying a car, they think of the size of the monthly payment, rather than the amount they will ultimately pay.

Teaching Notes

The net present value formula is difficult to derive before pre-calculus, but students can manipulate the formula, as long as they know how to use exponents. The formula is as follows:

$$A = R \,(1 - (1 + i)^n) \,/\, i$$

A is the present value of the loan (purchase price of the car), R is the monthly payments, i is the monthly interest rate (annual rate divided by 12), and n is the number of months in the loan period (for a 5-year auto loan, $n = 60$).

Extension Activity

Students can approach these exercises as potential car buyers. Local banks and credit unions will discuss terms and potential for credit, and often are willing to send a representative to your classroom.

Answers

1. Kahiye can afford to look for a car costing $6454.17 or less.

2. Bethany will pay $120.40 per month for her car, totaling $4334.40.

3. Thinh can afford to offer up to $7316.74 for the Civic.

4. Ethan can afford to pay up to $11,666.29 for his car.

5. Kaiulani will have to pay $234.38 per month for her car.

6. Megan's payments will be $369.15 each month. Her total payments (including her down payment) will be $24,649.

22. Auto Loan Rates

So you think you want to buy a car? Cars are expensive. In addition to the purchase price, you have to pay for maintenance, gas, insurance, and perhaps storage or parking. The size of your monthly payment is an important consideration in buying a car, but so are the total amount you are paying to use credit, and ensuring that you never owe more than the car is worth. The formula is as follows:

$$A = R (1 - (1 + i)^n) / i$$

A is the present value of the loan (purchase price of the car), R is the monthly payments, i is the monthly interest rate (annual rate divided by 12), and n is the number of months in the loan period (for a 5-year auto loan, $n = 60$).

Answer the following questions about auto loans that might arise in buying a car.

1. Kahiye Hassan needs a car to get to school every day. He has calculated that he can afford a payment of $160 each month. He is looking for a used car, and his credit union will give him a used-car loan for 4 years at an 8.8% annual interest rate. What is the most that Kahiye can afford to pay for a car?

2. Bethany Ives wants a car that she can use to commute to her horseback-riding lessons. She sees a 2000 Hyundai advertised for $3800. Her bank offers her a 3-year loan at 8.75%. What will her payments be each month, and how much will she pay for the car altogether over the 3 years?

3. Thinh Nguyen just noticed a 1999 Honda Civic for sale on a lot near his house. He can afford $180 per month, and the best loan he can get is 8.4% on a 4-year loan. What is the maximum he can afford to offer for the car?

(continued)

22. Auto Loan Rates

4. Ethan Chase needs a car to drive cross-country to get to college. He calculates that he can afford $225 per month, and a dealer is offering a 5-year loan at 5.9%. What is the most Ethan can afford to pay for a car?

5. Kaiulani Paine recently had her car totaled in an accident, so she needs a new one. She finds a 2002 Mazda Protege that she loves, and it costs $9750. The dealer is offering a 7.2% interest rate on a 4-year loan. What will Kaiulani's monthly payment be?

6. Megan Butler really wants the lime green 2006 Volkswagen Beetle she just saw at the dealership. The car costs $21,239. There is a 5% sales tax. Megan has $2500 for a down payment and will have to borrow the rest. If she can get a 4.5% interest rate on a 5-year loan, what will her payments be? Including her down payment and all 60 of her monthly payments, how much will Megan have paid for the car when she is done?

23. A Trip to Canada

Goal

To use conversion factors in switching between length and volume measures in the English and metric measurement

Context

The better part of the world's population uses metric measurement, while the United States and Britain continue to use the English measurement system. Many students traveling outside the country are faced with these conversions.

Teaching Notes

The conversion factors between English and metric measurements are listed in the activity. For students who are already familiar with conversion, the conversion factors should be enough to get them going on this activity. For students new to conversions, a discussion of multiplying fractions and conversions would be in order.

Extension Activity

Students might enjoy putting themselves in the place of a Canadian traveling to the United States. What conversions would they face?

Answers

Answers may vary slightly.

1. It is 366 miles from Montreal to Toronto, so it will take about 6 hours and 39 minutes.

2. The home run flew 394 feet.

3. Jasmin should keep her speed at 43 miles per hour.

4. The temperature in Windsor is predicted to be about 54°F.

5. She should buy the $\frac{1}{2}$ liter; it is still more than Jasmin usually drinks (about 16.9 ounces).

6. Windsor to Sudbury is about 476 miles. No, the trip is more than Jasmin's projected 400 miles per day, so she won't make it.

7. Jasmin is paying $2.38 (U.S.) per gallon.

Name _____   Date _____

23. A Trip to Canada

Jasmin lives in the United States and is planning a trip to Canada. The problem is that in Canada, everything is measured using the metric system. She will need to contend with different numbers for speed limits, distances, gasoline fill-ups, temperatures, and container sizes. See if you can help Jasmin on her trip. Use the conversion factors below to help with your calculations.

Conversion Factors

English Length	Metric Length
1 inch	2.54 centimeters
.394 inches	1 centimeter
1 foot	30.48 centimeters
.033 feet	1 centimeter
1 yard	9144 meters
1.094 yards	1 meter
1 mile	1.609 kilometers
.622 miles	1 kilometer

Volume	
33.8 ounces	1 liter
1 ounce	.03 liters
1 gallon	3.785 liters

Temperature

Fahrenheit	Celsius
Fahrenheit degrees	$\frac{9}{5} \times$ (Celsius degrees) + 32

1. According to Jasmin's map, the distance from Montreal to Toronto is 588 kilometers. Jasmin needs to know approximately how long the drive will take her. Judging from the road type on the map, she figures she can average about 55 miles per hour. How long should the trip take?

(continued)

23. A Trip to Canada

2. Jasmin goes to a baseball game in Toronto. She notices that the distances to the outfield fences are listed in meters. A player hits a home run that travels 120 meters. How many feet did it fly?

3. Jasmin knows that speed limits are listed in kilometers per hour. The speedometer on her car has only miles per hour. When the speed limit in Canada is 70 kilometers per hour, what speed should Jasmin go in miles per hour?

4. Jasmin sees on the weather report in Toronto that the weather in Windsor tomorrow is expected to be 12°C. What temperature is that in degrees Fahrenheit?

5. Jasmin wants to buy a soft drink at a truck stop. Soft drinks are sold in $^1/_2$-liter, 1-liter, and 1 $^1/_2$-liter cups. If Jasmin is usually satisfied with a 12-ounce can of soda, which cup should she buy?

6. The trip from Windsor to Sudbury, Ontario is 765 kilometers. Jasmin has set a goal of never traveling more than 400 miles per day. Can she make the trip in one day?

7. Jasmin stops to buy gas and sees that the price is $0.74 (Canadian) per liter. If $1.00 (Canadian) is worth $0.85 (U.S.), how much is Jasmin paying (in U.S. $) per gallon of gas?

24. Hat Sizes

Goal

To use ratios to create a chart of the relationship between hat sizes and head measurements

Context

Clothing measurements, pants, for example, mention inseam length and waist circumference, and are understandable. Other sizing is less obvious. Many of us wear hats; some of them are adjustable, and some are fitted. The chart on this page could be used in determining hat size.

Teaching Notes

Hat sizes seem mysterious because they have no obvious connection to the measurements of the head. In fact, the size is determined by a ratio of the circumference of the head in inches divided by 3.14 (π). The formulas for hat size are: hat size = circumference in inches ÷ 3.14 and hat size = circumference in centimeters ÷ 7.98.

Extension Activities

Ask students to measure each other's or family members' heads and determine their hat size. Also, ask students to speculate why the formula is what it is. (Early haberdashers used circular forms on which to build hats. Was it easier to identify them by their diameter than by their circumference?)

Answers

Answers may vary slightly.

Hat size	Circ. inches in decimals	Circumference in inches	Circumference in centimeters
$6\,^3/_4$	21.2	$21\,^1/_5$ inches	53.8 cm
$6\,^7/_8$	21.6	$21\,^3/_5$ inches	54.8 cm
7	22	22 inches	55.8 cm
$7\,^1/_8$	22.4	$22\,^2/_5$ inches	56.8 cm
$7\,^1/_4$	22.8	$22\,^4/_5$ inches	57.8 cm
$7\,^3/_8$	23.2	$23\,^1/_5$ inches	58.8 cm
$7\,^1/_2$	23.6	$23\,^3/_5$ inches	59.8 cm
$7\,^5/_8$	24	24 inches	60.8 cm
$7\,^3/_4$	24.3	$24\,^3/_{10}$ inches	61.8 cm
$7\,^7/_8$	24.7	$24\,^7/_{10}$ inches	62.8 cm
8	25.1	$25\,^1/_{10}$ inches	63.8 cm
$8\,^1/_8$	25.5	$25\,^1/_2$ inches	64.8 cm
$8\,^1/_4$	25.9	$25\,^9/_{10}$ inches	65.8 cm
$8\,^3/_8$	26.3	$26\,^3/_{10}$ inches	66.8 cm
$8\,^1/_2$	26.7	$26\,^7/_{10}$ inches	67.8 cm

© 2007 Walch Publishing

24. Hat Sizes

Many hats are adjustable, but cowboy hats, top hats, dress hats, and fitted baseball caps are still sized. Hat size is determined by taking the circumference of the head in inches and dividing by *pi* (π = 3.14). The conversion of inches to centimeters is 1 inch = 2.54 centimeters. Using these two pieces of information, hat size can be calculated by dividing circumference in centimeters by 7.98. Hat sizes increase in eighths.

The employees at The Mad Hatter on Main Street have lost their sizing chart and need a new one for reference. Help them by completing the chart below.

Hat size	Circumference inches in decimals	Circumference in inches	Circumference in centimeters
$6\frac{3}{4}$			
	21.6		54.8 cm
		22 inches	
$7\frac{1}{8}$			
		$22\frac{4}{5}$ inches	
			58.8 cm
$7\frac{1}{2}$			
		24 inches	
			61.8 cm
$7\frac{7}{8}$	24.7		
		$25\frac{1}{10}$ inches	
			64.8 cm
$8\frac{1}{4}$			
		$26\frac{3}{10}$ inches	
			67.8 cm

25. The Musical Scale

Goal

To learn how to work with ratios in the context of the musical scale and string lengths

Context

The ratio of the lengths of the strings is the way in which many pianos and other stringed instruments are tuned. Other factors include the thickness of the string and the tension on the string. Students who own stringed instruments such as guitars and violins will find that these other methods are used for tuning.

Teaching Notes

The ratio of string lengths (using strings of constant thickness and makeup) is such that one octave is a ratio of 2:1. The third and fifth steps of the scale are in ratio to the first by 4:3 and 8:5, respectively. Students should measure carefully and maintain constant tension in order to get the proper sound from plucked strings.

Extension Activities

Encourage students to use clamps and a string of constant thickness and makeup to prepare a stringed instrument capable of playing a major chord. Students might also examine a grand piano; although piano strings are not of constant weight and thickness, the middle octaves are made up of similar types of strings.

Answers

1. The strings should be 42 inches and 21 inches in length.

2. The strings should be cut 21 inches, 28 inches, and $33\,^3/_5$ inches in length.

3. The high C string is 10 inches long.

4. The low C string is $21\,^3/_4$ inches long.

5. The E string in that octave is 30 inches long.

6. The strings should be 15 inches, 20 inches, 24 inches, and 30 inches long, respectively.

7. The other strings should be $5\,^1/_3$ inches, $6\,^2/_5$ inches, 8 inches, $10\,^2/_3$ inches, $12\,^4/_5$ inches, and 16 inches.

© 2007 Walch Publishing

25. The Musical Scale

The Pythagoreans, a group of Greek mathematicians and followers of Pythagoras (c. 582–c. 500 B.C.E.), discovered much of what we know about building our musical scale. Using a piece of thread or string, they could hold the string tight and pluck it to make sounds. They found that if they cut a string 3 cubits long into strings of 1 cubit and 2 cubits exactly, the sound was pleasing to the ear. We refer to this length ratio of 2:1 as an octave. In a C major scale, the notes are C, D, E, F, G, A, B, C, and the major chord is C, E, G, C. The Pythagoreans found that the string for G needed to be in ratio to high C as 4:3, and that the length of the E string needed to be in ratio to high C as 8:5.

What lengths should the following strings be to make well-tuned instruments? Remember, the shortest string will make the highest sound.

1. A string 63 inches in length is to be cut to make strings of one octave difference in sound. How long should each string be?

2. You have a string 42 inches in length that makes a pleasing sound. How long should you cut other strings to make this the low note in a major chord?

3. The E string on an instrument is 16 inches long. How long is the high C string?

(continued)

25. The Musical Scale

4.   The G string on an instrument is $14\frac{1}{2}$ inches long. How long is the low C string?

5.   The G string on another instrument is 25 inches long. How long is the E string in that octave?

6.   You want to make a four-stringed instrument that will play a major chord. You have a string 89 inches long. How long should each string be to make your instrument?

7.   You want to make a seven-stringed, major chord, two-octave instrument. The shortest (highest) string is 4 inches long. How long should the other strings be?

26. The Respiratory Quotient

Goal

To learn how to work with ratios in the context of physiology and metabolic gas exchange known as the respiratory quotient

Context

The respiratory quotient (R.Q.) is calculated by taking the number of CO_2 molecules produced in a metabolic gas exchange and dividing by the number of O_2 molecules consumed to produce that reaction. A rating of 1 (the least oxidation required) comes from a 100% carbohydrate. A 100% fat requires 23 O_2 molecules to produce 16 CO_2 molecules, and therefore has an R.Q. of .696.

Teaching Notes

The chemical makeup of carbohydrates is $C_6H_{12}O_6$. By adding 6 O_2 molecules, the body creates 6 CO_2 (carbon dioxide) molecules and 6 H_2O (water) molecules. The respiratory quotient then is 6 (CO_2 molecules) ÷ 6 (O_2 molecules) = 1.00.

Extension Activity

Students are generally very interested in discussions of nutrition and exercise. Finding a physiologist or nutritionist to come in and discuss these issues can give you many other areas for mathematical discussion. Your school's chemistry teacher may also feel qualified to discuss them.

Answers

1. Pure carbohydrates have an R.Q. of 1.00.

2. Palmitic acid has an R.Q. of 0.696.

3. The protein albumin oxidizes with an R.Q. of 0.818.

4. a. 0.768 liters of CO_2 have been produced.

    b. 3.6 – 0.768 liters = 2.832 liters of CO_2 were produced by converting nonproteins.

    c. 0.96 liters of O_2 have been consumed.

    d. 4.3 – 0.96 liters = 3.34 liters of O_2 were consumed by converting nonproteins.

    e. The nonprotein R.Q. is 2.832 ÷ 3.34 = 0.848.

Name _____ Date _____

26. The Respiratory Quotient

Kevin Woodhouse is an athlete. He is very concerned with the physiological processes that take place in his body to convert foods to energy. One of these processes is represented by the respiratory quotient (R.Q.), a measure of how much oxygen (O_2) is used in converting food to energy. As this metabolic gas exchange takes place, carbon dioxide (CO_2) is produced. The higher the quotient, the higher the carbohydrate content of the food. A lower R.Q. means that a food has a higher fat content. The R.Q. is calculated by taking the number of CO_2 molecules produced in a metabolic gas exchange and dividing by the number of O_2 molecules consumed to produce that reaction.

Consider the following information about R.Q. that Kevin has found and answer the questions.

1. The chemical makeup of carbohydrates is $C_6H_{12}O_6$. By adding 6 O_2 molecules to a carbohydrate molecule, the body creates 6 CO_2 molecules and 6 H_2O (water) molecules. What is the respiratory quotient of this exchange?

2. The chemical makeup of palmitic acid is $C_{16}H_{32}O_2$. By adding 23 O_2 molecules to a palmitic acid molecule, the body creates 16 CO_2 molecules and 16 H_2O molecules. What is the respiratory quotient of this exchange?

3. When the body burns proteins, carbon dioxide and water are not the only by-products. The protein albumin $C_{72}H_{112}N_2O_{22}S$ oxidizes with 77 O_2 molecules to produce sulfur trioxide (SO_3), 9 molecules of urea ($CO(NH_2)_2$), 63 molecules of CO_2, and 38 molecules of H_2O. What is the R.Q. of this oxidation?

(continued)

26. The Respiratory Quotient

4. Most diets are a mixture of carbohydrates, fats, and proteins. However, it is possible to determine the CO_2 produced and the O_2 used in burning proteins by measuring the nitrogen excreted by the body: 1 gram of nitrogen excreted indicates 4.8 liters of carbon dioxide produced by burning proteins. Similarly, 1 gram of nitrogen produced indicates 6.0 liters of oxygen consumed by burning proteins.

   a. If a resting body excretes .16 grams of nitrogen in a 20-minute period, how many liters of CO_2 have been produced by burning proteins?

   b. By analyzing the expired air coming from this body, it is found that 3.6 liters of CO_2 were produced. How much was the result of burning nonproteins?

   c. In the same test subject, how many liters of O_2 have been consumed by burning proteins?

   d. By analyzing the expired air, it is also found that 4.3 liters of O_2 were consumed. How much was the result of consumption by nonproteins?

   e. What is the nonprotein R.Q. of this scenario?

27. Delivering Heating Oil

Goal

To recognize and practice the use of ratios in a setting such as an oil company's delivery office

Context

Many oil companies have "automatic delivery" customers, to whom they agree to deliver oil on a regular basis all winter. To save on costs, the company wants to deliver as much oil as possible as infrequently as possible. More frequent deliveries of lesser amounts cost the company money that they can't charge customers. However, if they wait too long, the customer will run out of oil (and heat) and likely stop doing business with them. Delivery is based on a calculation of the k-factor of the home and heating degree days. Each time a delivery is made, the k-factor of the house is calculated. The k-factor and the number of recent degree days are then used to determine the next delivery date.

Teaching Notes

The k-factor of a house is the number of degree days since the last delivery divided by the number of gallons delivered. Multiply the k-factor by 176 to get the number of degree days before the ideal next delivery.

Extension Activity

Ask students to research average numbers of degree days for different times of year, and to calculate the k-factor for their own homes.

Answers

Answers may vary slightly.

1. The k-factor for 844 Belden Avenue is 4.8. Steve should order delivery after about 845 degree days.

2. The k-factor for 5465 South Dorchester is 4.7. Steve should order delivery after about 827 degree days.

3. 181 gallons were delivered to the Kletziens' house on February 5.

4. The k-factor for the Normans' house on Terry Hill Road is 5.1. Steve should order delivery after about 898 degree days.

5. The k-factor for the house at 9810 Kensington Parkway is 5.1. Steve should order delivery after about 898 degree days.

27. Delivering Heating Oil

Steve Amato is a dispatcher at Wilson Oil. He needs to know when his customers will need oil so that he can arrange to have it delivered that day. He figures out when to deliver oil based on heating degree days and k-factors. Each day Steve calls Dave Lakeland (his weatherman), who tells him how many heating degree days were recorded the previous day, and he adds it to his running total. The k-factor of a house is the number of degree days since the last delivery divided by the number of gallons delivered. Ideally, Steve would like a driver to deliver 176 gallons of oil per delivery. He determines the ideal delivery date by multiplying the k-factor by 176. Calculate the following information for each house below so Steve can plan the next delivery for them.

1. The house at 844 Belden last had a delivery on January 31. For the last delivery period, 168.4 gallons of oil were delivered after 808 degree days. What is the k-factor for 844 Belden Avenue? How many degree days should Steve wait for the next delivery?

2. The house at 5465 South Dorchester last had a delivery on January 17. For the last delivery period, 178.2 gallons of oil were delivered after 833 degree days. What is the k-factor for 5465 South Dorchester? How many degree days should Steve wait for the next delivery?

(continued)

27. Delivering Heating Oil

3. The last delivery to the Kletziens' house at 319 South Chester Road was made on
    February 5 after 825 degree days. The computer has recorded a k-factor of 4.55, but
    the receipt has been lost for the number of gallons delivered. To the nearest gallon,
    how many gallons were delivered to the Kletziens on February 5?

4. The last delivery to the Normans' house on Terry Hill Road was made on February 14;
    159.3 gallons were delivered. In that delivery period, there were 812 degree days.
    What is the k-factor for the Normans' house? How many degree days until they
    should receive a delivery?

5. The last delivery to the house at 9810 Kensington Parkway was made on December 9.
    For the last delivery period, 165.8 gallons of oil were delivered after 846 degree days.
    What is the k-factor for 9810 Kensington Parkway? How many degree days should
    Steve wait for the next delivery?

28. Packaging and Product Size

Goal

To learn about ratios of sides, area, and volume by working with packaging and advertising in retail settings

Context

Packages have become trademarks in our advertising age. If a company's bottle is shaped a certain way and they want to sell a larger bottle, they will probably simply expand the dimensions of the bottle without changing the shape and style. Because small changes in linear size can lead to large changes in volume, it is good to understand the ratios involved.

Teaching Notes

Given that R_s is the ratio of the sides, R_a is the ratio of the area, and R_v is the ratio of the volumes, the relationship between the sides of two 2-D objects and their areas is $(R_s)^2 = R_a$. Similarly, the relationship between the sides of two 3-D objects and their volumes is $(R_s)^3 = R_v$. Consequently, a $3 \times 3 \times 3$ cube contains 27 cubic units of volume, and a $4 \times 4 \times 4$ cube contains 64 cubic units of volume, more than twice as much! Students may need to use manipulatives in building cubes to see this relationship at work.

Extension Activity

Students can go to stores and find examples of larger and smaller packages of the same shape and do an analysis of the volume or area ratios between the packages.

Answers

Answers may vary slightly.

1. A 12-inch pizza is 113 square inches, and a 14-inch pizza is 153.9 square inches. A 12-inch should cost just less than 1.5 (1.44) times a 10-inch, and a 14-inch should cost just less than twice (1.96) a 10-inch.

2. The volume of the can would increase by a factor of 1.73, yielding a $20\,^3/_4$-ounce can.

3. The bars now measure 4.7 inches by 1.4 inches by .94 inches.

4. The new cans would measure 6.41 inches tall and 2.14 inches in diameter.

5. The small pizzas cost 4.7 cents per square inch, and the extra large costs 3.9 cents per square inch (a better deal). However, if your party has different tastes, they might prefer separate pizzas with different toppings.

(continued)

28. Packaging and Product Size

We tend to judge products by their comparative sizes, but sometimes we don't think enough about dimensions. A can that is twice as tall doesn't necessarily hold twice as much. We can make better judgments by comparing the ratio of an object's dimensions to its area or volume. The relationship between the sides of two 2-dimensional objects and their areas can be expressed as $(R_s)^2 = R_a$. The relationship between the sides of two 3-dimensional objects and their volumes is $(R_s)^3 = R_v$. R_s is the ratio of the sides, R_a is the ratio of the areas, and R_v is the ratio of the volumes. Below are several questions referring to packaging and size of products. See if your estimates are accurate.

1. Have you ever wondered why there seems to be so little difference in the sizes of 10-inch, 12-inch, and 14-inch pizzas, yet the prices are so different? Calculate the areas of those pizzas and you will see the difference. The number of inches used to describe a pizza is the diameter of the pizza. The area of the pizza is calculated by taking half of the diameter, squaring it and multiplying by *pi* (π = 3.14). So a 10-inch pizza has an area of 25π or about $78\frac{1}{2}$ square inches. How many square inches do 12-inch and 14-inch pizzas have? What is the ratio of areas for these three pizzas (these would also be the appropriate ratios of prices)?

2. A 12-ounce soda can is approximately $2\frac{1}{2}$ inches in diameter and 5 inches in height. In order to maintain the same shape of the can while enlarging it, the company would have to add twice as many inches to the height as to the diameter. If they added $\frac{1}{2}$ inch to the diameter and 1 inch to the height, by what factor would they increase the volume?

(continued)

28. Packaging and Product Size

3. Several years ago, candy bar makers figured out that consumers would find it more acceptable if they decreased the size of the bars rather than increasing the price. A bar 5 inches long, 1.5 inches wide, and 1 inch thick was decreased from 6 ounces to 5 ounces. What were the new dimensions of the bar?

4. To save money, a company that cans olives has decided to decrease the size of their cans from 16 ounces to 10 ounces. Because of the distinctive shape of their cans (which are $7\frac{1}{2}$ inches tall, but only $2\frac{1}{2}$ inches in diameter), they have decided to maintain the shape but shrink them proportionally. What will the dimensions of the new cans be?

5. Louie's pizzeria is running a special this month—three small one-topping pizzas for $10.99. A small pizza is 10 inches in diameter. Louie's also sells an extra-large pizza, which is an 18-inch pizza with one topping for $9.99. Which order for pizza is a better deal per square inch? Why might you prefer the other option?

83 *Real-Life Math: Fractions, Ratios, and Rates*

29. Price-Earnings Ratio

Goal

To learn about ratios in the context of price-earning ratios of stocks and how they are used in investment strategies

Context

If you read the stock quotes in the *Wall Street Journal* or a comparable publication, there is a great deal of information included about each stock. In one column is the price-earnings ratio, which is used as part of several stock investment strategies. If the P-E ratio is low, then earnings are high in comparison with the price of the stock. This can mean that the stock is a good deal, or that investor interest is low and the stock price is stagnant.

Teaching Notes

The price-earnings ratio of a stock is listed in stock reports in the newspaper each day. The ratio is calculated by taking the closing price per share and dividing it by the reported annual earnings of the company per share of stock.

Extension Activity

The stock quotations are a wonderful source of a variety of comparative and informative data. Students might be asked to follow a stock with a low P-E ratio and another with a high P-E ratio, and see how the prices of those stocks change over a 6- or 8-week period.

Answers

1. IBM is reporting earnings of about $6.17 per share.

2. J.P. Morgan is reporting earnings of about $7.23 per share.

3. MGM Grand has a P-E ratio of 19.

4. UnumProvident has a P-E ratio of 20.

5. Disney stock is selling at 106.93 per share.

6. McDonald's stock is selling at $47\frac{1}{4}$ per share.

7. Delta Air Lines has a P-E ratio of 9.

8. Texas Instruments is reporting earnings of about $4.55 per share.

29. Price-Earnings Ratio

Anthony Lanni is a personal investment advisor. He sits down with individuals and families and helps them plan their budgeting, spending, saving, and investing for future goals and retirement. One type of investment that Anthony has recommended over the years is stocks on the New York Stock Exchange and NASDAQ market. His investment analysts use any number of indicators to predict which stocks will go up in value. One of those predictors is the price-earnings ratio (P-E). This ratio is calculated by taking the closing price per share and dividing it by the reported annual earnings of the company per share of stock. When the ratio of the price per share of a stock over the earnings per share of the company is low, the analysts think it means that investor interest in the stock is low and it will not change much in price. On the other hand, if the P-E ratio is high, the stock may be overvalued.

Answer the following questions about price-earnings ratios.

1. IBM stock is selling for $98\,^3/_4$ per share and has a price-earnings ratio of 16. Approximately what are their earnings per share?

2. J.P. Morgan stock is selling at $101\,^3/_{16}$ per share and has a P-E ratio of 14. Approximately what are their earnings per share?

3. MGM Grand stock is selling at $35\,^{13}/_{16}$ per share and reports earnings of $1.88 per share. What is their P-E ratio?

4. UnumProvident stock is selling at $48\,^5/_8$ per share and reporting earnings of $2.43 per share. What is their P-E ratio?

(continued)

29. Price-Earnings Ratio

5.   Disney is reporting earnings of $2.89 per share. Their P-E ratio is 37. What is the approximate price of a share of Disney stock?

6.   McDonald's has a P-E ratio of 20 and is reporting earnings of $2.36 per share. What is the approximate price of a share of McDonald's stock?

7.   Delta Air Lines is reporting earnings of $12.68 per share. Their stock is selling for $114\,^1/_8$ per share. What is their P-E ratio?

8.   Texas Instruments stock is selling for $54\,^5/_8$ per share. The P-E ratio is 12. Approximately what are their reported earnings per share?

Share Your Bright Ideas

We want to hear from you!

Your name_____Date_____

School name_____

School address_____

City _____State _____Zip_____Phone number (_____)_____

Grade level(s) taught_____Subject area(s) taught_____

Where did you purchase this publication?_____

In what month do you purchase a majority of your supplements?_____

What moneys were used to purchase this product?

____School supplemental budget   ____Federal/state funding   ____Personal

Please "grade" this Walch publication in the following areas:

Quality of service you received when purchasing .... A   B   C   D

Ease of use... A   B   C   D

Quality of content.. A   B   C   D

Page layout .... A   B   C   D

Organization of material .... A   B   C   D

Suitability for grade level .... A   B   C   D

Instructional value... A   B   C   D

COMMENTS:_____

_____

What specific supplemental materials would help you meet your current—or future—instructional needs?

_____

Have you used other Walch publications? If so, which ones?_____

May we use your comments in upcoming communications?   ____Yes   ____No

Please **FAX** this completed form to **888-991-5755**, or mail it to

   Customer Service, Walch Publishing, P. O. Box 658, Portland, ME 04104-0658

We will send you a **FREE GIFT** in appreciation of your feedback. **THANK YOU!**